PING Junior Conference (PING 2020)

Modern Trends in Material Engineering

IOP Conference Series: Materials Science and Engineering
Volume 1161

Pilsen, Czech Republic
8 - 10 September 2020

Editors:

Ludmila Kucerova
Hana Jirkova
Stepan Jenicek

ISBN: 978-1-7138-3428-1
ISSN: 1757-8981

Printed from e-media with permission by:

Curran Associates, Inc.
57 Morehouse Lane
Red Hook, NY 12571

Some format issues inherent in the e-media version may also appear in this print version.

Institute of Physics
Dirac House, Temple Back
Bristol BS1 6BE UK

Phone: 44 1 17 929 7481
Fax: 44 1 17 920 0979

techtracking@iop.org

Additional copies of this publication are available from:

Curran Associates, Inc.
57 Morehouse Lane
Red Hook, NY 12571 USA
Phone: 845-758-0400
Fax: 845-758-2633
Email: curran@proceedings.com
Web: www.proceedings.com

TABLE OF CONTENTS

IOP Conf. Series: Materials Science and Engineering 1161 (2021) 011001 doi:10.1088/1757-899X/1161/1/011001

Preface

We would like to introduce you to the Proceedings of the 5th PING Junior conference - Modern Trends in Materials Engineering, which took place from 15 to 17 September 2020 in Pilsen, Czech Republic. The conference is regularly organized by the Regional Institute of Technology (RTI, www.rti.zcu.cz) at the University of West Bohemia.
This year's event was focused mainly on master and doctoral students in order to support their research activities and enable them to present the obtained results. Furthermore, selected research projects were presented within the contributions.

This year was very demanding for organizing mass events, and our conference was also affected by the coronavirus epidemic. For this reason, it was organized in a combination of full-time and online participation.

A total of 18 contributions were presented at the conferences, which reflected the research projects solved within RTI. Several presentations were devoted to high-strength steels, such as the optimization of their heat treatment, use for hot stamping technology or their corrosion properties. Other contributions were focused on optimizing the heat treatment of stainless steel. Presentations, which were concentrated on heat or cryogenic treatment of maraging steel in the conventional state and after 3D printing, were also presented. Attention was also paid to special technologies such as semi-solid processing. Of these, six were selected for publication in the proceedings of IOP Conference Series.

In conclusion, we would like to invite you to the sixth year of the PING conference, which will take place from 7 to 10 September 2021.

Organizing committee of PING 2020

IOP Conf. Series: Materials Science and Engineering 1161 (2021) 011001 doi:10.1088/1757-899X/1161/1/011001

Scientific Committee

Doc. Ing. Ludmila Kučerová, Ph.D., RTI, University of West Bohemia, Czech Republic
Ing. Štěpán Jeníček, RTI, University of West Bohemia, Czech Republic
Dr.-Ing. Hana Jirková, Ph.D., RTI, University of West Bohemia, Czech Republic

Organizing Committee

Ing. Štěpán Jeníček, RTI, University of West Bohemia, Czech Republic
Doc. Ing. Ludmila Kučerová, Ph.D., RTI, University of West Bohemia, Czech Republic
Ing. Kateřina Opatová, RTI, University of West Bohemia, Czech Republic
Ing. Michal Peković, RTI, University of West Bohemia, Czech Republic
Ing. Jiří Vrtáček, RTI, University of West Bohemia, Czech Republic
Ing. Kateřina Rubešová, RTI, University of West Bohemia, Czech Republic

Program Committee

Ing. Štěpán Jeníček, RTI, University of West Bohemia, Czech Republic
Doc. Ing. Ludmila Kučerová, Ph.D., RTI, University of West Bohemia, Czech Republic

Contacts

University of West Bohemia
Faculty of Mechanical Engineering – Regional Technological Institute – PING Conference
Univerzitní 8
306 14 Plzeň
Czech Republic
Phone: +420 377 638 762
E-mail: jeniceks@rti.zcu.cz

www.ping.zcu.cz

Proceedings PING Junior 2020 - Modern trends in material engineering

Editors: Doc. Ing. Ludmila Kučerová, Ph.D.
Dr.-Ing. Hana Jirková, Ph.D.
Ing. Štěpán Jeníček

Published by:
University of West Bohemia
Univerzitní 2732/8, 301 00 Pilsen, Czech Republic, Europe

Pilsen 2020

PING Junior 2020 is organized with the support of funds for specific university research project SVK1-2020-005

This conference is organized under the patronage of the Regional Institute of Technology of the University of West Bohemia in Pilsen.

Peer review statement

All papers published in this volume of *IOP Conference Series: Materials Science and Engineering* have been peer reviewed through processes administered by the proceedings Editors. Reviews were conducted by expert referees to the professional and scientific standards expected of a proceedings journal published by IOP Publishing.

- **Type of peer review**: Single blind
- **Conference submission management system:** conference websites: http://www.ping.zcu.cz/en/zaslat_prispevek.html
- **Number of submissions received:** 15
- **Number of submissions sent for review**: 10
- **Number of submissions accepted:** 6
- **Acceptance Rate (Number of Submissions Accepted / Number of Submissions Received X 100):** 60%
- **Average number of reviews per paper:** 2,3
- **Total number of reviewers involved: 14**
- **Any additional info on review process:**

  Step 1.
  Authors submit articles through online systems.
  The authors submitted the manuscript before the conference via the conference website (http://www.ping.zcu.cz/en ---> Submit a contribution, abstract).

  Step 2.
  The responsible editor checks the formalities: whether the article thematically corresponds to the conference, whether the correct template is used and whether the required number of pages is observed, etc. If the formalities are met, the editor-in-chief appoints 2 reviewers. Reviewers are given 2 weeks to prepare a review and send the review to the editor. Reviewers are sent a prescribed review form with areas and questions for review which includes the following sections:
  A. Professional and science level,
  B. Formal structure,
  C. Total evaluation of the paper and publishing.
  The editor nominates another reviewer if the reviewer's comment contradicts itself (1 reviewer accepted the article and 1 reviewer rejected it).

Step 3.
The responsible editor decides whether the reviewer's comment is relevant or not. If the comment is relevant, the reviewer's comment and the editor's comment will be sent to the author for revision of the article. If they are not relevant, the editor can judge the article himself or request another revision by another reviewer.

Step 4.
Authors are required to revise their contributions according to the points listed in the reviewers' comments and the editor's comments.

Step 5.
The editor then evaluates the revised version to incorporate the points raised by the reviewers. If the reviewer requested a new review of the edited article, it is sent to the original reviewer. If the editor approves a revised version of the manuscript, the paper will be accepted for publication. If not, the editor sends the edited document to the author for further revision.

Step 6.
If the revised version does not match the points raised by the reviewer and the editor, the editor will send a rejection email to the author.

Contact person for queries:

Name: Štěpán Jeníček
Affiliation: University of West Bohemia
Email : jeniceks@rti.zcu.cz

Multi Elements Simulation of Biaxial Test with Two Different Soil Layers Using Hypoplastic Constitutive Model

O Khalaj [1], S Abedin Nejad [2] and T Janda[1]

[1] Regional Technological Institute, University of West Bohemia, Univerzitní 22, 30614, Pilsen, Czech Republic

[2] Faculty of Civil, Water and Enviromental Engineering, Shahid Beheshti University, Tehran, Iran

khalaj@rti.zcu.cz

Abstract. Practically, the ground condition usually consists of several layers, which have various properties from each other so it is important to have precise understanding about these layers and their behaviour. For this purpose, two different soil layers in biaxial test was analysed. In the single element simulation, it is impossible to generate the shear strain localization in the porous media but we can get numerous useful results. For strain localization analyse, using multi element models is necessary. The numerical simulation shows that the behaviour of each layer is totally different and the friction angle of both layers changed during simulation.

1 Introduction

Geotechnical engineering involves lots of challenges which usually related to the variety of mechanical properties of different layers even in a small area. It means that behavior of soil in one point of the field may be different from other point even in the depth. We have many soils layer that their properties are different, properties differences are in both depth and surface. Consequently, it is necessary to investigates soil layers behavior specially in finite element. From continuum mechanics standpoint, another important aspect of this simulation is usage of constitutive models. Note that several aspects can affect simulation results, like boundary conditions, loading values, shape and size of sample, etc. Many researchers discussed about shape effects and other aspects on experimental results, but limited studies were done in numerical space that considered soil layers by modern constitutive models like Hypoplastic. Masin and Suchomel (2008), investigated horizontally stratified deposit of sandy soil by means probabilistic approach [1]. The mechanical behavior of soil is described by an advanced hypoplastic and basic Mohr-Coulomb constitutive models. They calibrated models by using an automate procedure base on a set of experimental data from quarry Willin a granular grid ("figure 1"). comparison of typical experimental and simulated q vs. ε_a and ε_v vs. ε_s curves are shown in “figure 2 and figure 3” (on specimens for each sampling level). The hypoplastic calibration, is by the outlined procedures reproduces closely the q vs. ε_a curves and it should be noticed that, the rate of dilatancy is underestimated.

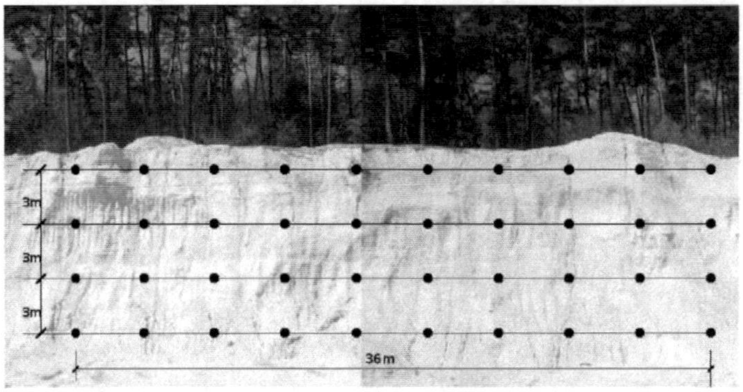

Figure 1. sand pit wall in Trebon basin. Positions of specimens are represented by black dots.

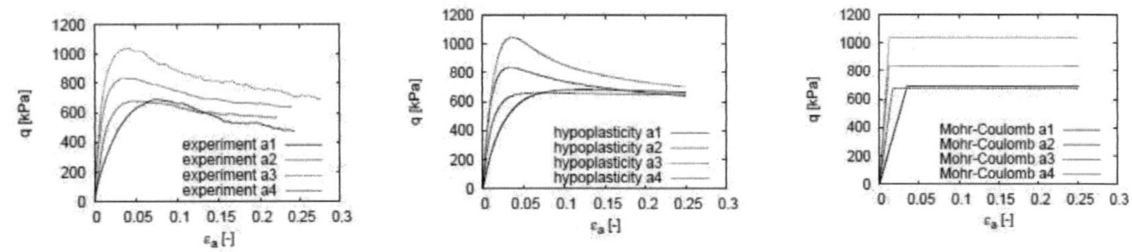

a b c

Figure 2. Typical experimental q vs. ε_a curves (a) compared with predictions by the hypoplastic (b) and Mohr-Coulomb (c) constitutive models, one specimen for each sampling level.

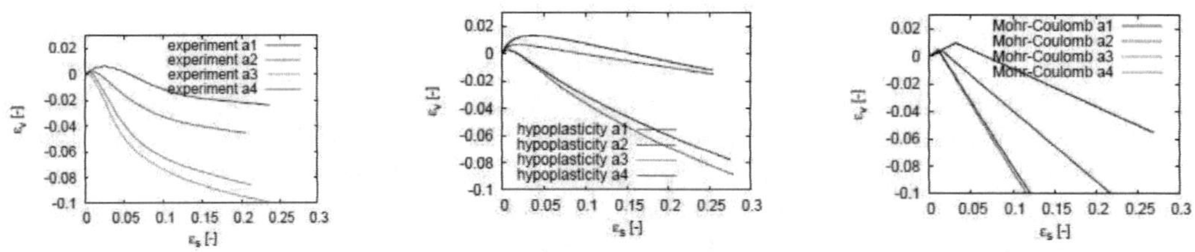

a b c

Figure 3. Typical experimental ε_v vs. ε_s curves (a) compared with predictions by the hypoplastic (b) and Mohr-Coulomb (c) constitutive models, one specimen for each sampling level

Tejchman and Wu (1996) studied about shear band formation with hypoplastic constitutive model in numerical simulations in Biaxial test [2]. They believed that start of shear bands formation is depends on initial materials property. "Figure 3" shows the calculations were done by the meshes that consist of the triangular using a linear shape function that each node is freedom in two degrees.

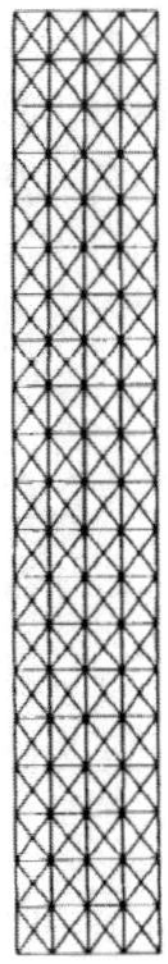

Figure 4. mesh used in calculations (plain strain)

For the first calculations, homogeneous mesh without any imperfection was considered. The sample was assumed to be dense. Load vs. vertical displacement curve is illustrated in "Figure 5" although, the deformation is inhomogeneous no shear zone can be observed. in the deformed shape of specimen. the deformation form is like a barrel because of self-weight effect. for $u_2 = 1$ cm, top and bottom boundaries have about 0.5 cm differences horizontally.

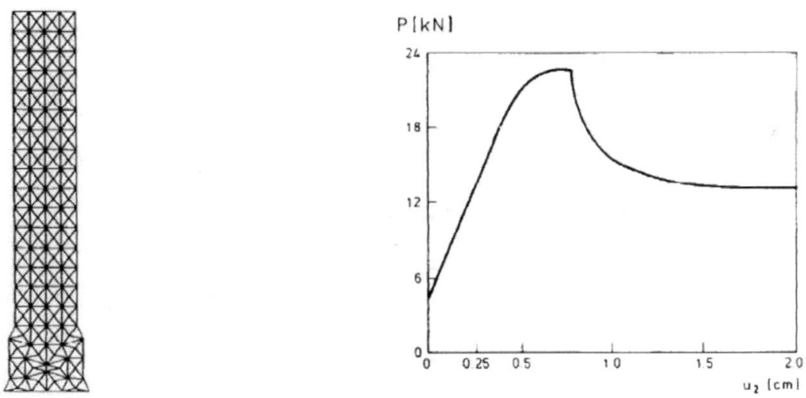

Figure 5. deformed mesh for $u_2 = 1$cm and load-displacement curve (without imperfection in mesh)

Mesh with an imperfection for dense sample, was analyzed and its results are shown in "Figure 6 – 8". When the shear starts, two different shear regions appear in the specimen and start to grow from weakness. This Growth is specified in Void ratio's distribution that is shown in "Figure 8, note that along the Hight in the middle of the specimen. Only one shear zone starts to develop after all the vertical displacement occurred. It is considered by concentration of displacements (fig 7) moreover in the shear localization region, void ratio increases (fig8) beyond the shear localization, the void ratio does not change very much. In hypoplastic constitutive model, the void ratio is a good criterion for shear areas. By characteristics length for extension the hypoplastic model, it is possible to describe the thickness of the shear zones. Desrues and Hammad (1989) showed shear banding is being influenced by stress level and the initial density, however, this consideration is being neglected in most of the studies and only one initial density under a given stress level is considered [3].

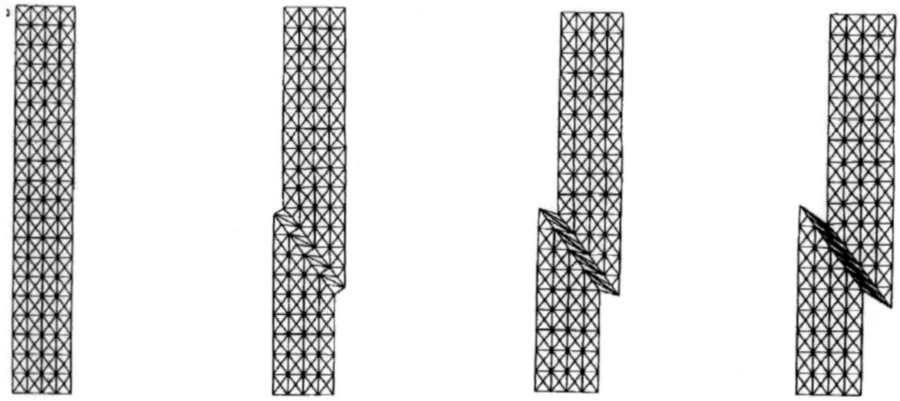

Figure 6. deformed mesh for with imperfection in mesh for (a) $u_2 = 0.6$cm (b) $u_2 = 1.2$cm (c) $u_2 = 1.6$cm (d) $u_2 = 2$cm

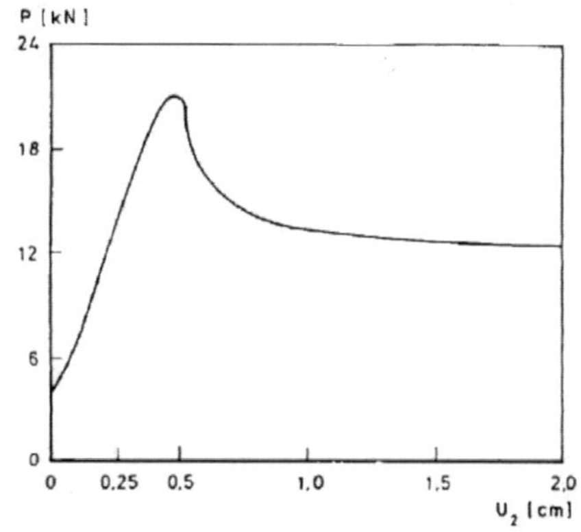

Figure 7. load-displacement curve with weakness in mesh

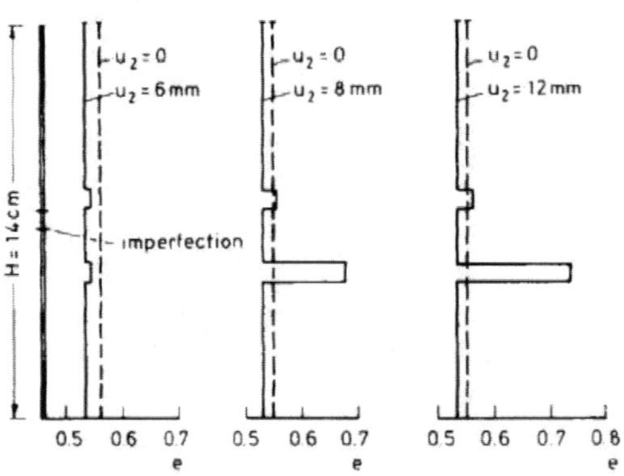

Figure 8. distribution of void ratio along the height in the middle of the specimen for different displacements u_2 of the top edge

2 Hypoplastic formulation

Hypoplasticity has numerous differences in compare with elastoplasticity theories. For example, unlike elastoplastic models, Hypoplastic without and yield surface or plastic protentional, that is common definition in elastoplastic constitutive models, is capable to predict inelastic and incremental non-linear material. The first hypoplastic constitutive theory was proposed by Kolymbas, Bauer, Gudehus and Niemunis as follow:

$$\dot{T} = F(T, e, D) \tag{1}$$

Where tensor function F depends on stress T, void ratio e and stretching rate D. For the determined stretching rate D, the formulation yields to stress rate response $\dot{T}$. the response function F can be written as [4, 5]:

$$F := f_s(trT, e)(L(\hat{T}, D) + f_d(trT, e)N(\hat{T})\|D\|) \tag{2}$$

Where in (2), L and $N\|D\|$ are the tensorial parts which D depends on the stretching rate and $\hat{T}$, depends on the stress ratio tensor. $\hat{T} = T/tr\ T$ is dimensionless and coaxial with T. The scalar quantities f_s, that is the pressure dependent stiffness factor and f_d is density factor. These factors take into account the influence of the mean skeleton pressure, $p_s = -(T_{s1} + 2T_{s2})/3$ and void ratio e, [4, 6, 7]
The tensorial equation can be divided into the fallowing two scalar expressions [5].

$$\dot{T}_{s1} = f_s \frac{(T_{s1} + 2T_{s2})^2}{T_{s1}^2 + 2T_{s2}^2} \left[D_1 + a^2 \frac{T_{s1}D_1 + 2T_{s2}D_2}{(T_{s1} + 2T_{s2})^2} T_{s1} + f_d \frac{a}{3} \frac{5T_{s1} - 2T_{s2}}{T_{s1} + 2T_{s2}} \sqrt{D_1^2 + 2D_2^2} \right] \tag{3}$$

$$\dot{T}_{s2} = f_s \frac{(T_{s1} + 2T_{s2})^2}{T_{s1}^2 + 2T_{s2}^2} \left[D_2 + a^2 \frac{T_{s1}D_1 + 2T_{s2}D_2}{(T_{s1} + 2T_{s2})^2} T_{s2} + f_d \frac{a}{3} \frac{4T_{s2} - T_{s1}}{T_{s1} + 2T_{s2}} \sqrt{D_1^2 + 2D_2^2} \right] \tag{4}$$

Where subscribes 1 and 2 denote the axial and radial directions, respectively:

$$f_s = \frac{h_s}{n} \left(\frac{e_i}{e}\right)^{\beta} \frac{1 + e_i}{e_i} \left(\frac{3p_s}{h_s}\right)^{1-n} \left[3 + a^2 - a\sqrt{3} \left(\frac{e_{i0} - e_{d0}}{e_{c0} - e_{d0}}\right)^{\alpha}\right]^{-1} \tag{5}$$

$$f_d = \left(\frac{e - e_d}{e_c - e_d}\right)^{\alpha} \tag{6}$$

$$a = \frac{\sqrt{3}(3 - \sin\phi_c)}{2\sqrt{2}\sin\phi_c} \tag{7}$$

Basic Hypoplastic model include the eight parameters $(\varphi_c, n, h_s, e_{i0}, e_{d0}, e_{c0}, \alpha, \beta)$. Void ratio parameters have indirect relations with p_s [6]; [4].

$$\frac{e_i}{e_{i0}} = \frac{e_c}{e_{c0}} = \frac{e_d}{e_{d0}} = \exp[-(\frac{3p_s}{h_s})^n] \tag{8}$$

2 Numerical modelling and analysis

One of the most important tools for simulation and numerical research is finite element software that in This paper we used ABAQUS finite element software, sample width is 8cm and height is 20cm. we assumed that our sample includes 2 different layers that both layers is analysed with basic Hypoplastic constitutive model. Boundary conditions In Biaxial compression test is demonstrated in "figure 9" and this simulation include 2 steps that in the first step cell pressure is applied at sides and top surfaces of sample. The amount of cell pressure is 200 kPa. After applying cell pressure, the deviatoric stress is applied in the top side of sample. For considering deviatoric stress, top boundary moved 5 cm along -y direction. Material properties of both layers is shown in table 1.

Table 1. material properties of soil layers

Parameters → / Soil ↓	φ_c [°]	h_s [GPa]	n [-]	e_{d0} [-]	e_{c0} [-]	e_{i0} [-]	α [-]	β [-]
Hochstetten sand	33	1.5	0.28	0.55	0.95	1.05	0.25	1.5
Toyoura sand	30	2.6	0.27	0.61	0.98	1.1	0.18	1.1

First simulation was done in multi elements that the bottom layer is Hochstetten sand and top layer is Touyora sand in second simulation, layers arrangement is reversed.

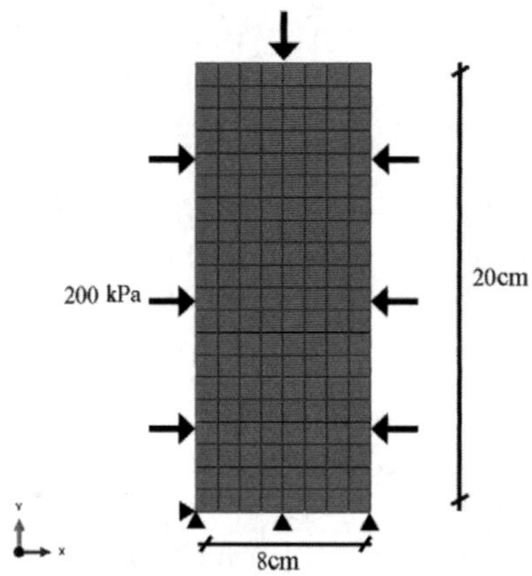

Figure 9. Sample dimension, boundary condition and loading status

It should also be added that due to the unstable nature of materials, it is necessary to consider pre-pressure for sample that value of this pressure is 1kPa, this instability generated due to lake of cohesion. Deformed shapes of two simulations are demonstrated in "figure 10. strain localization is occurred in both models. These results show that in both models strain localization created in Toyoura layer for both simulations. it shows that friction angle is the most important factor of behaviour of soils in Hypoplastic constitutive law. In "figure11" mobilized friction angle is demonstrated.

a b

Figure 10. deformed sample with their strain localization a) Top layer is Toyoura b) bottom layer is Toyoura

a b

Figure 11. mobilized friction angle a) Top layer is Toyoura b) bottom layer is Toyoura

In "figure 12" stress-Axial displacement curve for both simulations is shown. It shows that in both simulation ultimate stresses are same and both models stress-strain behaviors are so close. "figure 13" shows volumetric strain-axial strain curves that are close to each other.

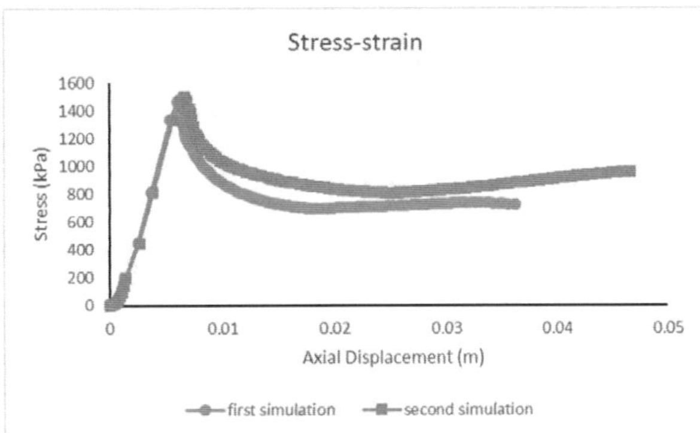

Figure 12. axial displacement a) Top layer is Toyoura b) bottom layer is Toyoura

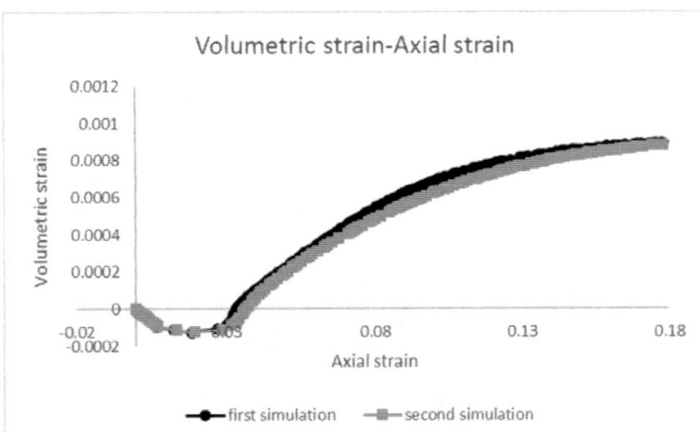

Figure 13. volumetric strain-Axial strain a) Top layer is Toyoura b) bottom layer is Toyoura

3 Conclusion

As discussed above, in some areas, soil may behave differently and engineers face several types of ground conditions that each layer has different properties. One of the most important tools for numerical investigation of soil behaviour is finite element software, as ABAQUS software used for this paper. Hypoplastic constitutive model was chosen to evaluate the material properties. Hypoplastic constitutive model are based on nonlinear tensor functions and are characterised by simple formulation and few parameters. Hypoplastic constitutive model can predicts soil behaviour more realistic than classical specially for granular materials. Two different models were used, Hochstetten sand and Toyoura sand are used in simulation and Properties of these soils are almost similar and just have only minor differences, but most of the strain localization is in the Toyoura sand and this phenomenon is due to differences between friction angle in Toyoura sand and Hochstetten sand, so in hypoplastic model friction angle parameter has significant effects on results and predictions compared with other parameters.

References

[1] Suchmoel, R. & Mašin D, Calibration of an advanced soil constitutive model for use in probabilistic numerical analysis, 2008
[2] Tejchman, J & Wu,W., numerical simulation of shear band formation with a hypoplastic constitutive model, computer and geotechnics, Vol.18, No. 1, pp. 71-84,1996
[3] Desrues, J. & Hammad,W., shear banding dependency on mean pressure level in sand. 2nd int. workshop on localisation and bifurcation theory for soils, Gdańsk, Poland 1989
[4] Bauer. E, 1996, Calibration of comprehensive Hypoplastic model for granular materials. Soils Found 36(1):13-26Roscoe KH, Schofield AN, Wort CP on the yielding of soils Géotechnique 8:22-53, 1958
[5] Von Wolffersdorf. PA, A Hypoplastic relation for granular materials with a predefined limit state granular surface. Mech Cohes-Frict Mater 1:25-271, 1996
[6] Bauer. E, Wu. W, A hypoplastic constitutive model for cohesive powders. Powders Tech 85:1-9, 1995
[7] Herel. I, Gudehus. G, Determination of parameters of hypoplastic constitutive model from properties of grain assemblies. Mech Cohes-Frict Mater 4(5):461-486, 1999
[8] Wu W (1992) Hypoplasticity as a mathematical model for the mechanical behaviour of granular materials. PhD thesis, Karlsruhe University, Germany

IOP Conf. Series: Materials Science and Engineering 1161 (2021) 012002 doi:10.1088/1757-899X/1161/1/012002

Effect of prior annealing on the microstructure and mechanical properties of thermo-mechanically treated TRIP steel

L Kucerova[1], K Burdova[1], A Stehlik[1], I Ticha[1]

[1]Laboratory of Metallography, Regional Technological Institute, University of West Bohemia in Pilsen, Univerzitní 22, 306 14, Plzeň, Czech Republic

E-mail: skal@rti.zcu.cz, kburdova@rti.zcu.cz, stehlika@rti.zcu.cz, ichenati@rti.zcu.cz

Abstract. Low-carbon low-alloyed high strength steel 0.2C-1.5Al-0.5Si-2Mn-0.06Nb (in weight %) was forged into bars and annealed at 1050 °C with either 2-hour hold and furnace cooling to room temperature or with 20-min hold and controlled cooling by 1 °C/s. These various holds and coolings created two different initial microstructures, banded ferritic-pearlitic one for the long hold and bainitic-ferritic for the shorter hold and quicker cooling. Samples with those microstructures then underwent the same thermo-mechanical treatment (TMT) typical for TRIP (transformation induced plasticity) steels. Very different microstructures and mechanical properties were obtained for both initial states even after the application of the same TMT schedules. Light and scanning electron microscopy was used to characterize obtained microstructures and tensile tests were applied for the determination of mechanical properties.

1 Introduction

TRIP (Transformation induced plasticity) steels are commonly low-carbon low-alloyed steels, alloyed only by small amounts of silicon or aluminium and manganese with possible micro-alloying by niobium, molybdenum or titanium. TRIP steels are commonly produced either by a combination of cold rolling with subsequent two-step heat treatment or by thermo-mechanical treatment [1-5]. This heat or thermo-mechanical treatment should result in multiphase microstructures with the potential to utilize a TRIP effect. For low-alloyed TRIP steel, the required microstructure would be either ferritic-bainitic or bainitic, in both cases with the controlled amount of retained austenite (typically 10-20%) [6-7]. Appropriate stability of retained austenite ensures its gradual transformation into martensite during cold plastic deformation which results in strengthening of the material at one side and prolonged uniform deformation stage during straining at the other side. If the austenite is not stable enough, it will transform suddenly during the small plastic deformation, while if it is over-stabilized, the transformation might not have occurred at all before the fracture. In both of those cases, the TRIP effect would not be triggered and mechanical properties of the steel would not be improved [8]. The same positive effect of retained austenite can be observed also in other grades of advanced high strength steels, such as in quenched and partitioned (QP) martensitic steels [9-11].

Low-alloyed TRIP steels are typically used for an automotive application and the final products are processed mainly from thin sheets [12-13]. Most of the researchers pay attention to the optimisation of dedicated heat or thermo-mechanical treatment of the product, using cast ingots, which are first re-forged or pre-rolled and annealed to provide input material for TRIP steel processing [14-18]. This work aims at the evaluation of the effect of different initial microstructure, created by different annealing treatments, on the final microstructure and mechanical properties after subsequent thermo-mechanical treatment.

2 Experimental program

Low-alloyed low-carbon steel 0.2C-1.5Al-0.5Si-2Mn-0.06Nb (in weight %) used in this work doesn't meet any standard and therefore was custom-cast. Cast ingots of experimental steel were cut into four equal parts and hot-forged into bars with a diameter around 25 mm and air-cooled to room temperature.

The first halves of the bars were annealed at 1050 °C for 2 hours in the furnace and then furnace-cooled to room temperature. The second halves were used to produce directly samples for the thermo-mechanical simulator [20] and therefore annealed at 1050 °C for 20 min and controlled cooled at 1 °C/s to the room temperature. The samples for the thermo-mechanical simulator had a cylindrical shape with a diameter of 8 mm and an active length of 16 mm which was optimized to ensure homogeneous distribution of heat and deformation across the active length of the sample

The microstructure obtained by slower furnace cooling consisted of bands of ferrite and pearlite, while the microstructure obtained by controlled cooling was more homogenous and was consisted of pearlitic and bainitic blocks surrounded by the chains of polygonal ferrite grains which were created at prior austenite grain boundaries (**Figure 1**). The hardness of both steels was also different, the banded microstructure obtained by 2-hour annealing had around 250 HV10 while the microstructure after shorter annealing possessed the hardness around 320 HV10.

The banding effect observed after slower cooling from 1050 °C in this work presents a generally known issue in the processing of medium Mn steels and was already described in hot rolled and 1000 °C solution annealed steel with nearly the same chemical composition Fe-0.17C-1.47Mn-1.48Si (in weight %) by Farahani et al. [19]. The formation of ferrite-pearlite bands in those steels is caused by significant partitioning of alloying elements during solidification which is hard to remove sufficiently during heat treatment. However, faster cooling from annealing temperatures is generally believed to suppress this effect. Moreover, Farahani et al. [19] also demonstrated this cooling rate dependency in a set of dilatometric experiments, finding out that the bands already formed at cooling rates of 1°C/s or slower. In the case of this work, the cooling rate of 1°C/s was sufficiently quick to prevent banding in used CMnAlNb steel.

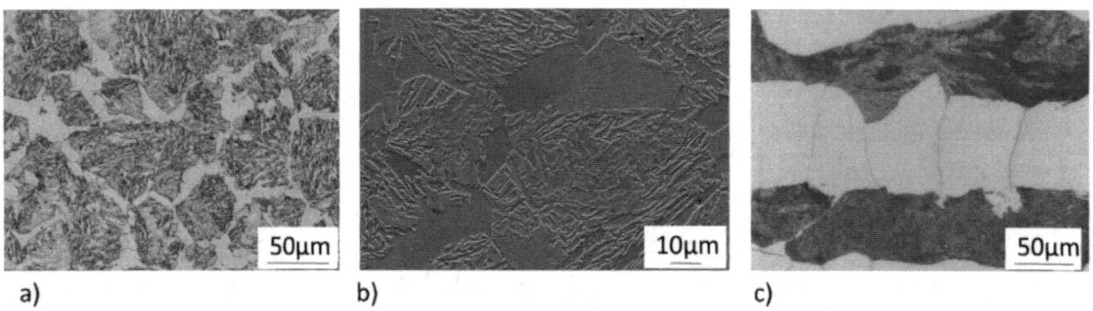

Figure 1. Initial microstructures after annealing at 1050 °C/20 min-1°C/s a),b) and annealing at 1050 °C/2 hours-furnace cooling to RT.

Subsequently, both types of microstructures underwent the same thermo-mechanical treatment with soaking temperature of 900 °C, which is in one phase (fully austenitic) region of used steel according to JMatPro calculations and this was also confirmed by previous experiments [18]. After 100 s hold at soaking temperature, the first 10% compression deformation was applied still at the temperature of 900 °C. The second compressive deformation was applied during the cooling at a temperature of 720 °C. The cooling rate was 30 °C/s, down to the isothermal hold. Three different temperatures of the hold were tested - 350, 450, and 500 °C. Based on the previous experiments with similar steels [3, 6], the holds of 1200 s and 600 s were chosen for the temperatures of 350 °C and 450 °C to assess the progress of bainite transformation. A significantly shorter hold of 25 s was applied at the highest temperature of 500 °C to prevent pearlite formation.

Microstructures were characterized in the central part of the longitudinal section of the processed samples by the means of BX61 Olympus light microscope and VEGA 3 and EVO 25 scanning electron microscopes (SEM). Mechanical properties were measured by a tensile test of flat samples with an active length of 5 mm [21] and a cross-section of 1.2 x 2 mm.

Results and discussion

All the microstructures obtained by TMT of samples annealed for 20 minutes with 1°C/s cooling to room temperatures were bainitic with a small amount of proeutectoid ferrite placed at prior austenite grain boundaries (**Figure 2**a-d). No banding and segregation were visible in those samples. On the other hand, all samples which were pre-treated by 2-hours annealing and slow furnace cooling possessed markedly banded microstructure, with bands of polygonal ferrite and bands of martensite or bainite (**Figure 2**e-h). The difference between the microstructures with various annealing conditions is clearly apparent from the comparison of light micrographs taken at lower magnification. Different microstructures are also reflected in different mechanical properties of the samples with the same TMT (**Table 1**). Samples with 1050 °C/2 hours annealing followed by furnace cooling always achieved higher strengths and lower elongation than corresponding samples with 1050 °C/20 min annealing followed by 1 °C/s cooling.

The sample with 1050 °C/2 hours annealing followed by furnace cooling and a short hold at the temperature of 500 °C (1050°C/2 hours-500°C/25 s) had mainly bainite in the bands (**Figure 3**e) and reached a high yield (576 MPa) and tensile strengths (1018 MPa) with lower elongation of 22%. Apparently, the sample with 1050 °C/20 min annealing didn't have any martensite or ferrite in the final microstructure after the same TMT. The microstructure was in this case mainly bainitic with prevailing granular bainite morphology (**Figure 3**a). This microstructure had a strength of 900 MPa and 33% of total elongation.

Table 1. Mechanical properties of samples yield tensile strength (YTS), ultimate tensile strength (UTS), total elongation (TE).

Annealing	Isothermal hold T/t [°C/s]	YTS [MPa]	UTS [MPa]	TE [%]
1050°C/2hours	500/25	576±5	1018±3	22±2
1050°C/20min	500/25	493±3	900±7	33±2
1050°C/2hours	450/1200	550±4	900±5	24±1
1050°C/20min	450/1200	549±5	819±1	45±0
1050°C/2hours	350/1200	785±2	1031±2	20±4
1050°C/20min	350/1200	653±1	880±6	30±0
1050°C/2hours	350/600	682±8	998±4	23±1
1050°C/20min	350/600	673±16	912±2	27±0

As in the previous case, similar microstructure development can be seen in the sample with 20 min hold at 450 °C (1050°C/120 min-450°C/20min) in which mainly bainite was found within the bands including large islands of martensitic-austenitic (M-A) constituent. This constituent typically forms in TRIP steels, when larger islands of remaining austenite are not stable enough and partially transform to martensite during the final cooling. Central parts of those larger islands tend to transform into martensite more easily, while the edges often still contain retained austenite. Bainitic blocks in this microstructure were rather coarse and very dense with little bainitic ferrite visible. However, as more bainite was obtained than in the case of 500 °C isothermal hold, the resulting tensile strength was only 900 MPa, while the total elongation did not increase significantly (24%). The sample with shorter annealing (1050°C/20 min-450°C/20 min) had again predominantly bainitic microstructure with a mixture of granular and lath bainite. Larger islands of retained austenite were found mainly at the edges of granular bainite blocks. In some cases the formation of M-A constituent was also visible.

This microstructure achieves the best elongation of 45% with the tensile strength of 819 MPa, which is a very good combination of mechanical properties for this type of steel.

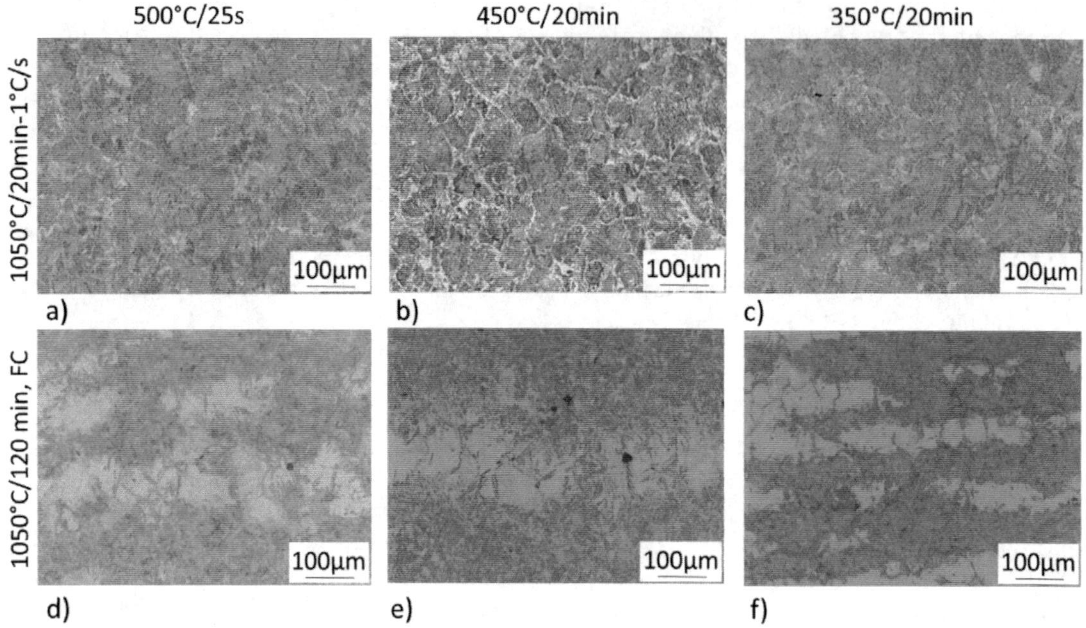

Figure 2. Light micrographs of samples: a) 1050°C/20 min-500°C/25 s, b) 1050°C/20 min-450°C/20 min, c) 1050°C/20 min-350°C/20 min , d) 1050°C/120 min-500°C/25s, e) 1050°C/120 min-450°C/20 min, f) 1050°C/120 min-350°C/20 min (FC – furnace cooling).

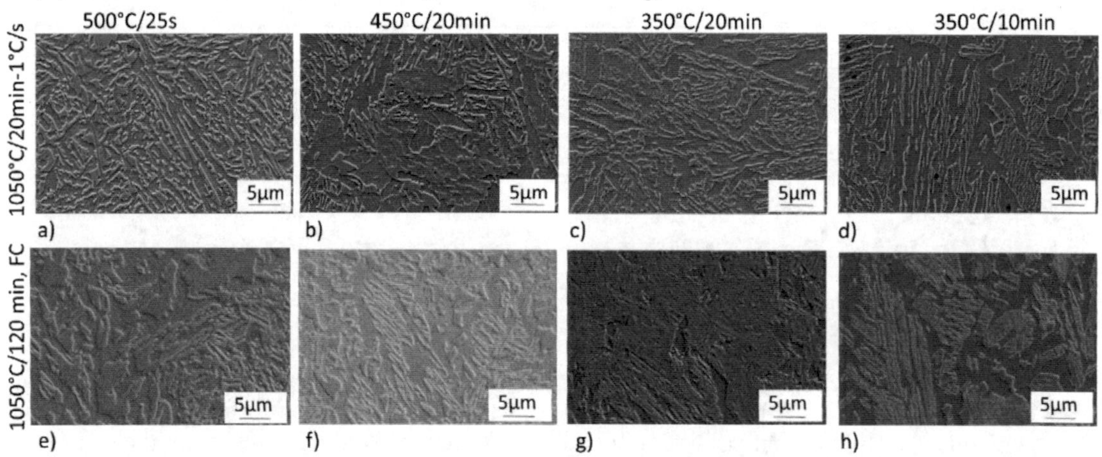

Figure 3. Scanning electron micrographs of the samples: a) 1050°C/20 min-500°C/25 s, b) 1050°C/20 min-450°C/20 min, c) 1050°C/20 min-350°C/20 min , d) 1050°C/20 min-350°C/10 min, e) 1050°C/20 min-500°C/25s, f) 1050°C/120 min-450°C/20 min, g) 1050°C/120 min-350°C/20 min, h) 1050°C/120 min-350°C/10 min (FC – furnace cooling).

The lowest isothermal hold of 350 °C also produced banded microstructure for both hold times for the samples with longer annealing at 1050 °C, consisting of polygonal ferrite and martensite-bainite bands (1050°C/2 hours-350°C/20min, 1050°C/2 hours-350°C/10min). The sample with 20 min hold at 350 °C had mainly very dense martensite-bainitic microstructure, containing areas of martensitic blocks with very little ferrite and areas of lath bainite which resulted in the increase of tensile strength to 1031 MPa and decrease of elongation to 20%.

The sample with shorter 10 min hold at 350 °C (1050°C/2 hours-350°C/10min) had a similar microstructure with rather coarser martensitic-bainitic blocks, some of the bainite reaching up to 50μm long bainitic laths. The islands of the M-A constituents are however larger and more frequent resulting in the slightly lower tensile strength of 998 MPa and higher elongation of 23%.

The samples with shorter annealing and isothermal hold of 350 °C (1050°C/20 min-350°C/20min, 1050°C/20 min-350°C/10min) had bainitic microstructure with occasional proeutectoid ferrite at prior austenite boundaries. However, larger areas of granular bainite were found in the sample with a shorter, 10 min hold, at 350 °C. This microstructure also possessed slightly higher yield (673 MPa) and tensile strength (912 MPa) and lower total elongation (27%) than the microstructure obtained by longer, 20 min hold at 350 °C. Longer hold at 350 °C resulted in the yield strength of 653 MPa, tensile strength of 880 MPa, and 30% total elongation.

Conclusions

Two sets of samples of low-alloyed low-carbon TRIP steel were prepared by forging and solution annealing. One set was annealed for two hours at 1050 °C and furnace cooled to the room temperature, while the other was annealed at the same temperature only for 20 minutes and then cooled at 1 °C/s. Totally different microstructures were obtained, longer hold, and slower cooling resulted in banded ferrite-martensite or ferrite-bainite microstructure and shorter hold and quicker cooling produced bainitic microstructures with a small fraction of proeutectoid ferrite. The same thermo-mechanical treatments were applied to both sets of samples.

Even though a fully austenitic state was achieved during the soaking, traces of the original microstructure were not completely erased and the banded original microstructure resulted in banded final microstructure after the thermo-mechanical processing. The same parameters of thermo-mechanical processing applied to the homogeneous original microstructure resulted in also the homogeneous distribution of the final microstructure. This difference was reflected by mechanical properties as well, the banded microstructures generally achieved higher strengths and lower total elongations than microstructures produced by the same processing parameters applied to homogeneous input microstructure. The highest tensile strength of 1031 MPa was achieved after solution annealing with a longer hold at 1050 °C and slow furnace cooling with subsequent thermo-mechanical treatment with 20 min hold at the temperature of 350 °C. The highest total elongation was on the other hand obtained for the sample with a shorter hold at 1050 °C and controlled cooling at 1 °C/s, followed by thermo-mechanical treatment with 20 min hold at 450 °C.

Acknowledgement

The present contribution has been prepared with the support of the student grant competition of University of West Bohemia in Pilsen, SGS 2019-019. The project is subsidised from specific resources of the state budget for research and development.

References

[1] Alirez M S and Mirzadeh K H 2020 Transformation-induced plasticity (TRIP) in advanced steels A review, Materials Science and Engineering: A, 795 140023.

[2] Pereloma E V, Timokhina I B, Hodgson P D 1999 Transformation Behaviour in Thermomechanically Processed C–Mn–Si TRIP Steels with and without Nb Mater. Sci. Eng. A. 273–275 448–452

[3] Kucerova L, Opatova K, Kana J 2017 High versatility of Niobium alloyed AHSS Archives of

Metallurgy and Materials 62(3) 1485-1491

[4] Monsalve A, Guzmán A, De Barbieri F 2016 Mechanical and Microstructural Characterization of an Aluminum Bearing Trip Steel Metallurgical and Materials Transactions A 47(6)

[5] Shen Y F, Qiu, L N, Sun, X 2015 Effects of retained austenite volume fraction, morphology, and carbon content on strength and ductility of nanostructured TRIP-assisted steels Mater. Sci. Eng. A. 636 551-564

[6] Kučerová L, Jirková H, Mašek B. 2014 The Effect of Alloying on Mechanical Properties of Advanced High Strength Steels. Archives of Metallurgy and Materials 3(59) 1189-1192

[7] Li L, De Cooman B C, Wollants P, He Y, Zhou X 2004 Effect of Aluminum and Silicon on Trans-formation Induced Plasticity of the TRIP Steel, J. Mater. Sci. Technol. 20 135-138

[8] Timokhina I B, Enomoto M, Miller M K, Pereloma E V 2012 Microstructure-Property Relationship in the Thermomechanically Processed C-Mn-Si-Nb-Al-(Mo) Transformation-Induced Plasticity Steels Before and After Prestraining and Bake Hardening Treatment Metallurgical and Materials Transactions A 43A(7), 2473-2483

[9] Kucerova L, Aisman D, Jirkova H 2009 Optimization of Q-P process parameters with regard to final microstructure and properties, Annals of DAAAM for 2009 & proc. of the 20th international DAAAM symposium (Annals of DAAAM and Proceedings) 20 1035-1036

[10] Hauserová D, Duchek M, Dlouhý J, Nový Z, Properties of Advanced Experimental CMnSiMo Steel Achieved by QP Process, Procedia Engineering, 10, 2961-2966

[11] Kučerová L, Jirková H, Káňa J 2016 The suitability of 42SiCr steel for Quenching and Partitioning Process. Manufacturing Technology. 16(5) 984-989

[12] Hong S Ch, Lim S H, Hong H S, Lee K J, Shin D H Lee K S 2003 Effects of Nb on strain induced ferrite transformation in C-Mn steel Mater. Sci. Eng. A. 355 241-248

[13] Xiaodong Tan, Huansheng He, Wenjun Lu, Liu Yang, Di Wu 2020 Effect of matrix structures on TRIP effect and mechanical properties of low-C low-Si Al-added hot-rolled TRIP steels, Materials Science and Engineering: A, 771 138629

[14] Chen Shih-Che et al. 2020 Role of the crystallographic texture in anisotropic mechanical properties of a newly-developed hot-rolled TRIP steel, Materials Science and Engineering: A, 790 139683

[15] Mašek B, Jirková H, Kučerová L, Zemko M. 2011 Improvement of Mechanical Properties of automotive Components Using Hot Stamping with Integrated Q-P Process. Journal of Iron and Steel Research International, 18 (Supplement 1-2) 730-734

[16] Li Z C et al. 2018 The significant impact of pre-strain on the structure-mechanical properties relationship in cold-rolled medium manganese TRIP steel, Materials Science and Engineering: A, 712 206-213

[17] Kumar A, et al. 2007 Microstructural and mechanical characterization of C–Mn–Al–Si cold-rolled TRIP-aided steel, Materials Science and Engineering: A, 445-446 549-557

[18] Kučerová L, Jirková H, Volkmannová J, Vrtáček J 2018 Effect of Aluminium and Manganese Contents on the Microstructure Development of Forged and Annealed TRIP Steel. Manufacturing Technology, 18(4) 605-610

[19] Farahani H, Xu W. Zwaag S. Van Der 2018 A Novel Approach for Controlling the Band Formation in Medium Mn Steels. Metallurgical and Materials Transactions A, 49A 1998-2010

[20] Vorel I, Jeníček S, Jirková H, Mašek B 2016 Material and technological modelling of closed-die forging. Materiali in Tehnologije, 50(4), 499-503. doi:10.17222/mit.2014.220

[21] Kučerová L, Jeníček Š, Mach J 2018 Effect of sample geometry on tensile properties of chromium alloyed middle carbon steel. Paper presented at the METAL 2018 - 27th International Conference on Metallurgy and Materials, Conference Proceedings, 761-765

Experimental modelling of materials properties and microstructure of new high-strength steels for press-hardening and hot metal gas forming

Kateřina Opatová*, Štěpán Jeníček, Adam Stehlík, Hana Jirková

University of West Bohemia, RTI - Regional Technological Institute, Univerzitní 8, CZ-301 00 Pilsen, Czech Republic

*opatovak@rti.zcu.cz

Abstract. Press-hardening is a highly dynamic process that involves rapid temperature changes. These process aspects play a major role in microstructural evolution and mechanical properties. In order to develop any thermomechanical processing sequence, including press-hardening, materials models of the best possible accuracy are needed, if relevant results are to be obtained from FEM methods employed for designing metalworking processes. Therefore, relevant data on materials behaviour under real process conditions must be obtained experimentally. In one such experiment, as described here, PHS 1800 and PHS 2000 low-alloy sheets of steel were subjected to thermal exposure identical to the heating sequence for press-hardening, including the removal from the furnace and transfer to the tool. Heat treatment sequences were proposed which correspond to the thermal profiles in the actual process. The results were input into FE simulations of press-hardening. In the low-alloy steels, the selected heat treatment sequence led to strengths of more than 2000 MPa and elongation levels of approx. 9%.

Key words: Press-hardening, PHS 1800, PHS 2000, Material-technological modelling

1. Introduction

The automotive industry is one particular sector that increases the demands on the properties of formed sheet metal. In order to meet these demands, various methods are being developed, such as press-hardening. (Fig. 1). This article explores opportunities for altering the mechanical properties of final products and for accurately designing their final microstructures. Advanced methods of this kind might soon begin to bear on today's production of sheet metal parts for the automotive industry and may considerably improve these parts properties [1, 2]. Among the very promising materials, there are the PHS steels from the company SSAB which deliver strengths around 2000 MPa along with elongation levels of around 10 %. The treatment of these steels was designed under the effiPRESS project (RFSR – CT-2015-00019).

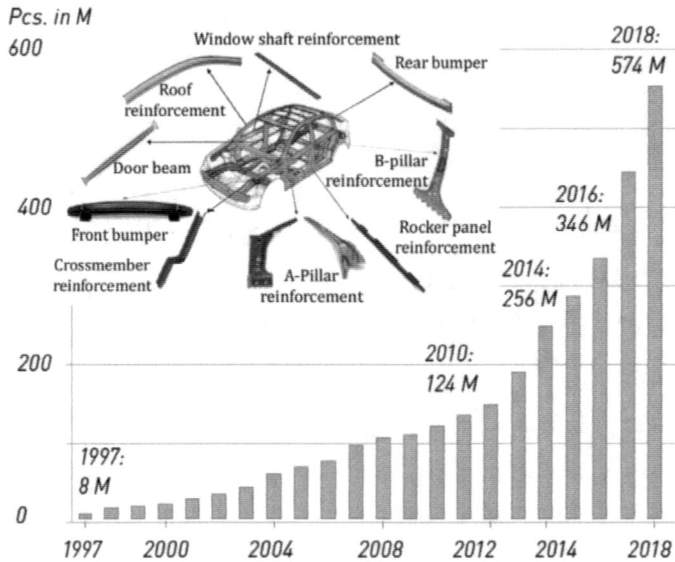

Figure 1: Number of press-hardened parts produced annually to meet demand [2, 3].

Press-hardening is a dynamic process in which the blank is heated and austenitized in a furnace and then transferred to a tool to be formed and quenched in a single step (Fig. 2) [10]. The blank is cooled rapidly by the contact with the tool. The deformation and cooling rates lead to hardening microstructures, providing the desired mechanical properties, namely high strength and sufficient ductility [4, 5]. Such properties are of major importance to the automotive industry because they lead to improved passenger safety and to potential weight savings and the related operation efficiency [6].

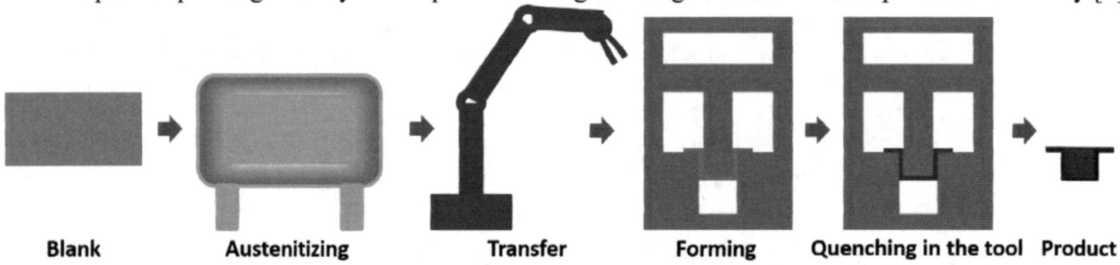

Figure 2: Schematic diagram of the press-hardening process.

2. Experimental Program

Two new UHS steels (ultra-high-strength) were specially designed for this experiment: PHS 1800 and PHS 2000. Despite their low alloy levels, they show potential for excellent mechanical properties (Tab. 1).

Table 1: Chemical composition of experimental materials declared by the manufacturer.

Composition [wt %]	C	Si	Mn	P	S	B
PHS 1800	0.27 – 0.33	Max 0.40	1.15 – 1.45	Max 0.025	Max 0.035	0.0008 – 0.0050
PHS 2000	0.33 – 0.40	Max 0.40	1.15 – 1.45	Max 0.025	Max 0.035	0.0008 – 0.0050

PHS 1800 and PHS 2000 were supplied in annealed and as-rolled condition, respectively. Their as-received microstructures consisted of ferrite and pearlite. The microstructure of as-received PHS

1800 showed signs of presence of globular cementite. By contrast, PHS 2000 contained lamellar pearlite and elongated ferrite grains which clearly indicated the rolling direction (Fig. 3). The as-received steels had distinct mechanical properties. The ultimate strength of PHS 1800 was 515 MPa, whereas that of PHS 2000 was 1070 MPa. In order to obtain comprehensive information on both materials, CCT diagrams were constructed using JMatPro software for designing the first processing sequences (Fig. 4).

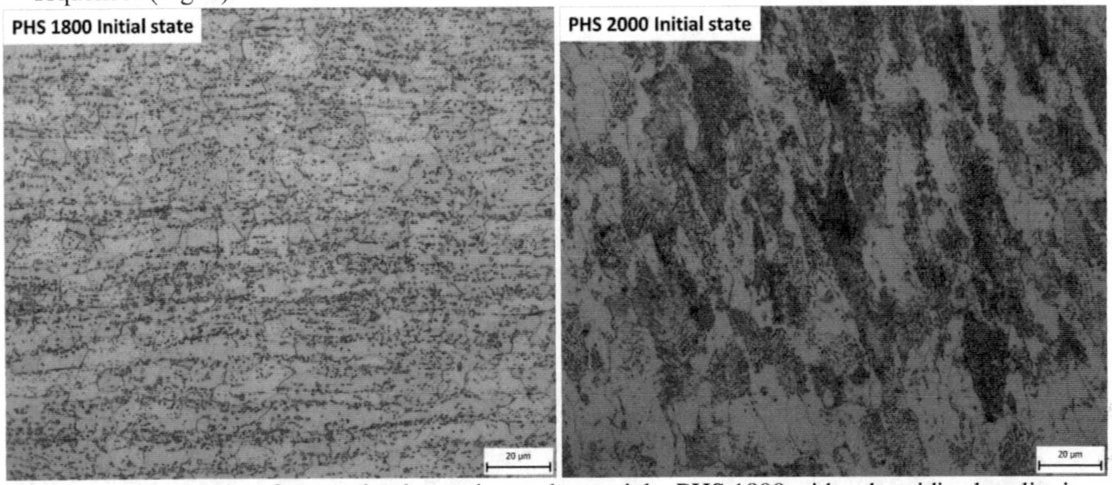

Figure 3: Micrographs of as-received experimental materials. PHS 1800 with spheroidized perlite in the ferritic matrix after annealing. The microstructure of PHS 2000 was ferritic-pearlitic with lamellar perlite after cold rolling.

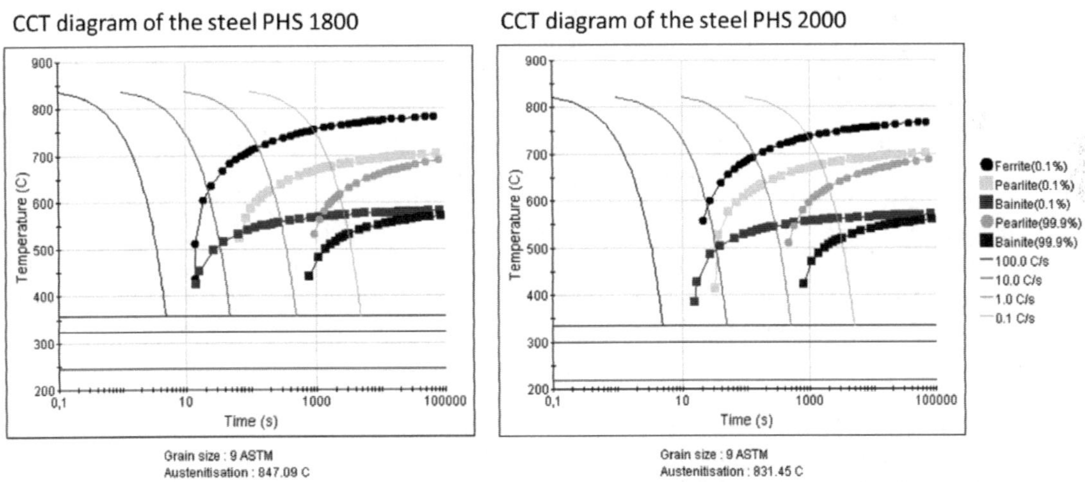

Figure 4: CCT diagrams of experimental materials constructed using JMatPro software (JMatPro, Release 9.0, Sente Software Ltd., 2016).

2.1. Data preparation for developing the material-technological model of the press-hardening process of experimental steels

For the purpose of characterizing the impact of the rolling direction upon mechanical properties and detecting any defects caused by rolling, the first step involved tensile testing of sheet metals in their initial condition in several directions (0°, 45°, 90°). In the next step, simplified press-hardening sequences were designed. These involved heating to 850°C, 900°C, 950°C and 1000°C and subsequent water and oil quenching. In these sequences, the holding times were varied. Between holding and quenching, a pause was made in order to simulate the real transfer from the furnace to the tool. This pause took between 7 and 12 seconds, depending on the specimen shape (sheet or tube) and the instrument (Fig. 5).

These sequences were carried out in a thermomechanical simulator. They provided a basis for identifying sufficient holding temperatures for both materials. Based on these sequences, parameters for simulations of the real press-hardening process were found by iteration. The purpose was to achieve the desired properties of the material within the shortest time and at the lowest possible costs.

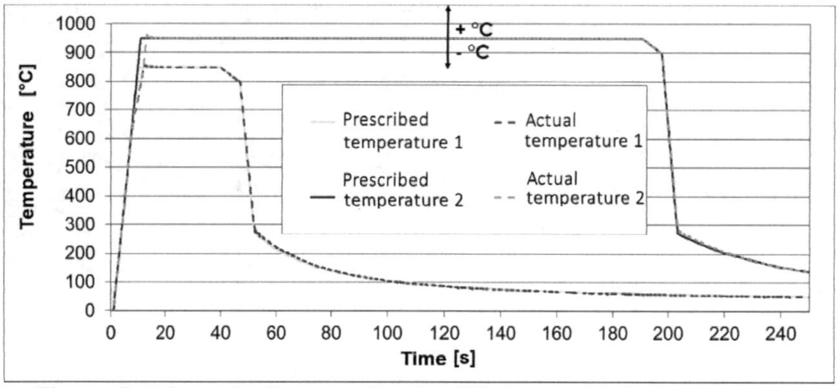

Figure 5: Schematic diagram of a simplified press-hardening sequence.

In PHS 1800, the temperature of 850°C was insufficient for full austenitization. The specimens contained ferrite and martensite. Although the extended holding time of 180 seconds – instead of 30 seconds – has indeed led to lower ferrite fractions, it was not sufficient for achieving a fully-martensitic structure. Full austenitization was only achieved by increasing the temperature to 900°C or more. Optimal mechanical properties were achieved with the austenitizing temperature of 900°C and isothermal holding for 180 s. After the austenitizing temperature was increased to 950°C and more, the ultimate strength decreased, most probably as a result of the onset of coarsening of the austenitic grain.

PHS 2000 became fully austenitized at a holding temperature as low as 850°C which was maintained for 30 s. In this steel, too, lower ultimate strengths of specimens were found when the holding temperature and time at temperature were increased (Fig. 7).

Higher holding temperatures led to lower strength characteristics in both steels (Fig. 6, 7). The likely cause is the on-set of grain coarsening. Considering the direction of rolling prior to quenching, it can be said that the results of tensile testing provided no proof of significant deviations in the case of individual rolling directions. The likely reason is the essence of the treatment of specimens which were fully austenitized before quenching.

2.2. Simulation of the real forming process

After evaluation and optimization of the parameters obtained from the simplified press-hardening process, parameters for preparing the material-technological model [11,13] of the real forming process were determined (Fig. 8). Specifically, this process was hot metal gas forming of tubes. Hot metal gas forming is used for shaping tubes by means of internal pressure of a working fluid [7, 12]. The resultant increase in the tube diameter of between 3% and 30% depends on the wall thickness and the steel type. Forming takes place in closed dies into which heated seamless or welded tube feedstock is placed. The advantage of this tube-forming process is that it does not require intricate-shaped dies and mandrels. The disadvantage is the working fluid pressure which may reach up to 800 bar [7, 8].

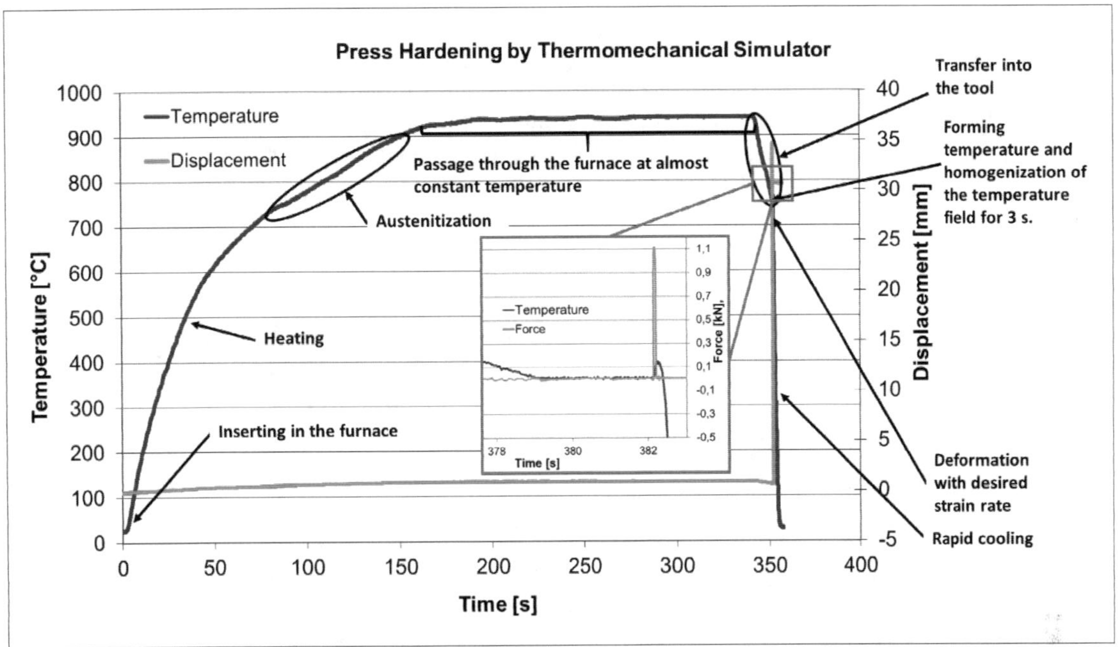

Figure 6: Schematic diagram of the simulation of the press-hardening process for tube forming.

Press-hardening is a highly dynamic process that involves rapid temperature changes. These process aspects play a major role in the microstructural evolution, and consequently in mechanical properties. In order to develop any thermomechanical processing sequence, including press-hardening, one needs materials models of the best possible ac-curacy, if relevant results are to be obtained from FEM methods employed for designing metalworking processes. Consequently, relevant data on materials behaviour under real process conditions must be obtained experimentally. For this reason, the steels under investigation were subjected to the same thermal schedule as the one applied to feedstock heating for press-hardening [6, 9].

The model consisted of heating the specimen to the desired temperature according to the actual heating curve. Two times at austenitizing temperature were used: 30 s and 180 s. This step was followed by a pause simulating the removal of the feedstock from the furnace and transfer into a tool. Afterward, deformation at various strain rates ranging from 0.5 to 50 s-1 was applied at various temperatures in the interval between 850 °C and 600 °C. Flow stress curves were obtained from these experiments. The flow stress curves which reflected the entire thermal exposure history for the press-hardening process were analysed and converted into true stress–true strain curves. Using numerical techniques, models for individual temperatures were constructed. This data became the basis for FE simulations of hot metal gas forming of welded tubes [5, 2].

3. Result and discussion

Information was obtained on the effects of austenitizing on final properties after quenching in oil and water. Oil-quenched PHS 1800 material had only become fully austenitized above 900°C. Lower processing temperatures cause proeutectoid ferrite to form, which considerably reduces strength properties. Optimum mechanical properties [14] were obtained with 900°C and isothermal holding time of 180 seconds in specimens with orientation parallel to the rolling direction – Rm = 1780 MPa and elongation A = 12%; in specimens perpendicular to the rolling direction – Rm = 1783 MPa and elongation A = 6%. Raising the austenitizing temperature above 900°C led to a decrease in strength (Fig. 7). Oil-quenched PHS 2000 material became fully austenitized above 850°C. In specimens that were parallel to the rolling direction, the maximum strength was 1950 MPa, along with elongation of A = 10% upon austenitizing at 850°C for 30 seconds. Similar mechanical properties profiles were obtained in specimens which were perpendicular to the direction of rolling (Rm = 1950 MPa, A = 8%)(Fig. 8). As with PHS 1800, raising the austenitizing temperature above this value led to a similar trend which involved decreasing ultimate strength. Mechanical properties of water-quenched specimens of both experimental steels were similar to those of the oil-quenched ones. From the process viewpoint, both steels can be quenched in oil or water without any appreciable impact on the outcome.

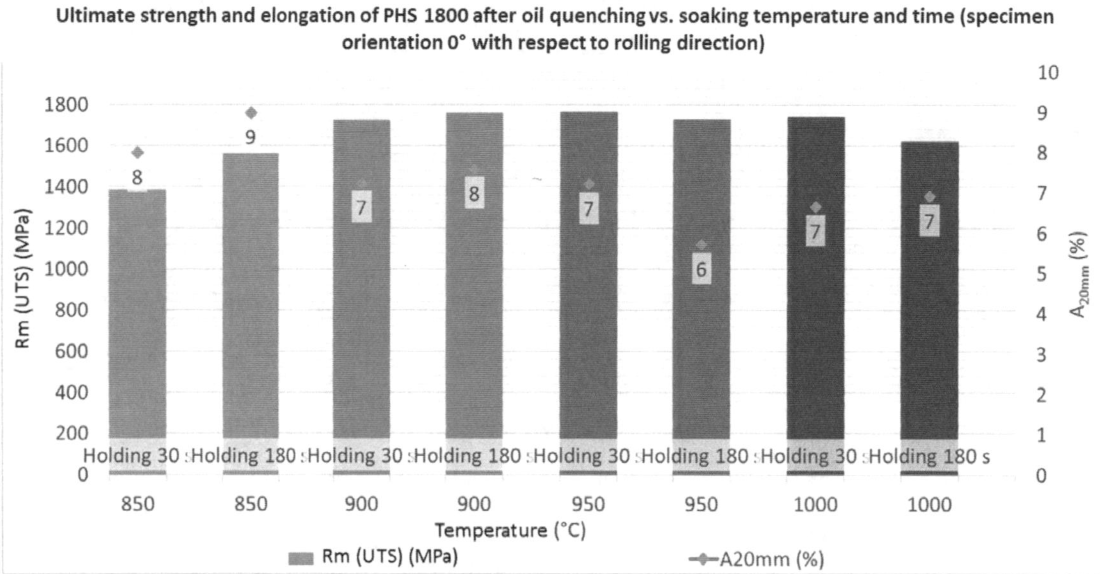

Figure 7: Mechanical properties of PHS 1800 after simplified press-hardening.

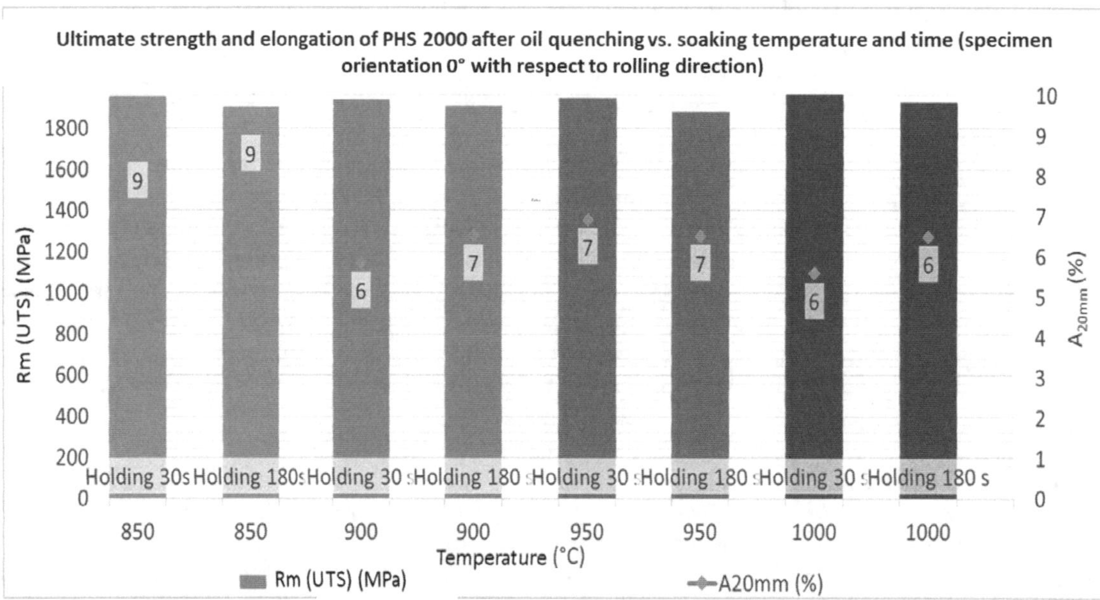

Figure 8: Mechanical properties of PHS 2000 after simplified press-hardening.

4. Conclusions

The research results achieved up to now suggest that it is possible to get highly accurate models of technological behaviour of sheet metals during hot forming under the conditions of the press-hardening process. This model includes all aspects of microstructure evolution associated with the thermal history of the material. Furthermore, experiments showed the behaviour of materials under the specific conditions of a hot forming process. This experiment incorporates the final rapid cooling in the tool. This rapid cooling is part of mostly used processes for the production of press-hardened parts. At the same time, information about the dependence of austenitization and austenitizing temperature on resulting properties after oil and water quenching has been obtained.

This collection of results enables a more sophisticated approach to be taken to designing various material processing routes involving press-hardening and other unconventional processes that rely on similar principles of formation of martensitic structures in thin-walled products of UHSS.

Acknowledgments

This paper includes results created within the project LO1502 Development of Regional Technological Institute. The project is subsidized by the Ministry of Education, Youth and Sports from specific resources of the state budget for research and development.

References

[1] Karbasian H, Tekkaya A E 2010 A review on hot stamping *Journal of Material science and Processing Technology* **210** p 2103

[2] Mori K, Bariani P F, Behrens B A, et. al 2017 Hot stamping of ultra-high strength steel parts, *CIRP Ann. - Manuf. Technol.* **66** p 755

[3] Schuler Group, Hot stamping with pch flex – fast, flexible, cost-effective, 2016. https://www.schulergroup.com/major/download_center/broschueren_hydraulic_press/download_hydraulic_press/hydraulic_press_leichtbau_broschuere_formhaerten_pch_e.pdf.

[4] Ruifeng D, Lifeng F A N, Xiaoyan Z 2016 Research on Stamping Forming Properties of the Steel Sheets Used for Car Body Proceedings of ICAS 2016 & HMnS 2016, p 603

[5] So H, Faßmann D, Hoffmann H, Golle R, Schaper M 2012 An investigation of the blanking process of the quenchable boron alloyed steel 22MnB5 before and after hot stamping process, *J. Mater. Process. Technol.* **212** p 437

[6] Jing G U O, Jing D, Guangdong Z, Zhiwei L I 2016 Research on Materials Substitute of Automobile Stamping Parts Based on the Simulation Proceedings of ICAS 2016 & HMnS 2016 p 656

[7] Vorel I, Jirková H, Mašek B, Kurka P 2015 Production of shaped semi-products from ahs steels by internal pressure *Mater. Tehnol.* **49** p 739

[8] Mašek B, Vorel I, Jirková H, Kurka P 2015 Combination of International High Pressure Forming and Q-P Process for Production of Hollow Products from AHS Steel *Adv. Mater. Res.* **1127** p 9

[9] Caron E, Daun K J, Wells M A 2013 Experimental characterization of heat transfer coefficients during hot forming die quenching of boron steel *Metall. Mater. Trans. B Process Metall. Mater. Process.* **44** p 332

[10] Jirková H, Opatová K, Jeníček Š, Vrtáček J, Kučerová L, Kurka P 2019 Use of multi-phase trip steel for press-hardening technology. Acta Metallurgica Slovaca, 25(2), 101-106. doi:10.12776/ams.v25i2.1267

[11] Ibrahim K, Vorel I, Jeníček S, Káňa J, Rubešová K, Opatová K, Kotěšovec V 2016 A study ofmaterial-technologicalmodelling for choosing the ideal cooling rate for designing production of closed die forgings using 30MNVS6 steel. Paper presented at the Annals of DAAAM and Proceedings of the International DAAAM Symposium, , 27(1) 551-555. doi:10.2507/27th.daaam.proceedings.081

[12] Jirková H, Jeníček Š, Kučerová L, Kurka P 2019 High-strength steel components produced by hot metal gas forming. Materials Science and Technology (United Kingdom), doi:10.1080/02670836.2019.1576355

[13] Vorel I, Jeníček S, Jirková H, Mašek B 2016 Material and technological modelling of closed-die forging. Materiali in Tehnologije, 50(4), 499-503. doi:10.17222/mit.2014.220

[14] Kučerová L, Jeníček Š, Mach J 2018 Effect of sample geometry on tensile properties of

chromium alloyed middle carbon steel. Paper presented at the METAL 2018 - 27th International Conference on Metallurgy and Materials, Conference Proceedings, 761-765.

IOP Conf. Series: Materials Science and Engineering 1161 (2021) 012004 doi:10.1088/1757-899X/1161/1/012004

Numerical Modelling of Round Bars Rolling Process in Three-High Skew Rolling Mill HDQT-R 30-12

M Peković, S Benešová

Faculty of Mechanical Engineering – Regional Technological Institute, University of West Bohemia, Univerzitní 8, 306 14 Pilsen, Czech Republic

E-mail: pekovicm@rti.zcu.cz, sbenesov@kmm.zcu.cz

Abstract: In the 21st century, numerical analysis is a proven tool for research and development of manufacturing processes. Mathematical modelling provides information about the behaviour of various materials during forming processes, for instance. The real-world process is therefore represented by an abstract mathematical model, on which calculations are performed. Three-roll skew rolling can be used for various grades of steel. This paper presents the results of a theoretical analysis of skew rolling of bars with an initial diameter of 30 mm for three different carbon steels, S235JRC, C45 and C60, and austenitic stainless steel AISI 321. The final diameter of the rolled bars was 22 mm and was achieved in a single pass. Several different rolling speeds were used for the stainless steel. The simulations were carried out using DEFORM 2D/3D software. The behaviour of the materials during rolling was analysed, including the values of effective strain and stress.

Keywords: FEM, Numerical analysis, Three-roll skew rolling, steel, round bars

1 Introduction

Trial and error or empirical approaches used to be the methods for designing some equipment for materials forming. However, the requirements for accuracy and reliability of products make it impossible to use them today [1]. In recent decades, advanced numerical modelling based on the finite element method has been widely used. In addition, these methods continue to be developed and have already found their place in many industries [2]. Using these computational methods, one can accurately predict the behaviour of the material during forming [1].

Several different processes can be used for making round bars. One of them is three-roll skew rolling. The range of materials includes aluminium, magnesium alloys, steel, titanium and other difficult-to-form materials. Three-roll skew rolling offers large reductions in a single pass along with high surface quality and roundness of the product. The products are less expensive to manufacture than the conventional-rolled products since there is no need for special rolls for different diameters [3]. Three-roll skew rolling is a complex process in terms of deformation profiles and stress states in the rolling gap. Many variables come into play and affect these states, including the rolling speed, the stock temperature, the type of feedstock, the product diameter and others [4, 5].

This paper presents an analysis of simulations of bar rolling performed in DEFORM 2D/3D software for HDQT-R 30-12 round bar forming equipment. The equipment is a three-roll skew rolling mill for input stock diameters of 30 mm and product diameter of 12 mm (Figure 1) [6]. The result of the simulations can be used as a basis for the design of new rolling processes, for example using other semi-finished products as an input.

Figure 1. The plant for rolling round bars HDQT-R 30-12 [7].

2 Materials and methods

The device for rolling round bars HDQT-R 30-12 (**H**igh **D**eformation **Q**uenching and **T**empering) can reduce the diameter of round bars. It is useful to be aware of the forces acting on the bar during rolling, the strain or stress distribution on its cross-section and the material flow in the process.

Using DEFORM 2D/3D software, the rolling of round bars with diameter reduction from 30 mm to 22 mm was simulated. These simulations were developed for four steels with different chemical compositions (Table 1). They were three carbon steels, C15, C45 and C60 with different carbon levels and austenitic stainless AISI 321.

Table 1. Chemical compositions of the steels used in the experiments.

Hm. %	C	Si	Mn	P	S	Ni	Cr	Mo	Ti
DIN C15	0.12-0.18	max 0.4	0.3-0.8	max 0.045	max 0.045	x	x	x	x
DIN C45	0.43-0.5	max 0.4	0.5-0.8	max 0.045	max 0.045	max 0.4	max 0.4	max 0.1	x
DIN C60	0.57-0.65	max 0.4	0.6-0.9	max 0.045	max 0.045	max 0.4	max 0.4	max 0.1	x
AISI 321	max 0.08	0.25-0.1	max 0.2	x	x	9-12	17-19	max 0.75	0.7

In all the simulations, the roll speed was 100 rpm. Additional simulations for AISI 321 involved roll speeds of 50 and 200 rpm. The temperature of the stock before rolling was 1100°C.

A three-dimensional model for the simulation had to be developed using CAD software based on the drawings supplied with the equipment. The model was transferred to the DEFORM and a finite element mesh was created which divided the stock into 193630 tetrahedra (Figure 2).

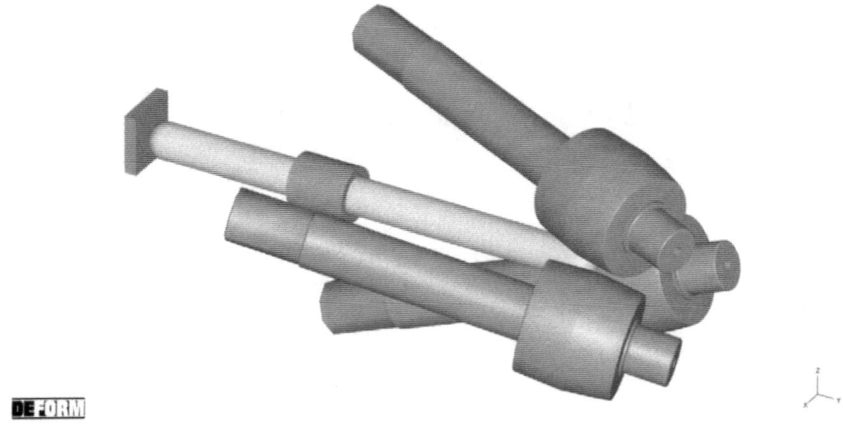

Figure 2. The default configuration for numerical simulations.

3 Results and discussion

In the simulations, material flow, stress distribution across the material cross section and effective deformation in the rolled product were observed.

3.1 Plastic flow during rolling

The material flow during rolling was explored using two features of the software. The first one, "flownet tracking" mapped the movement of virtual lines on the surface of the stock and revealed that they rotate during the rolling process (Figure 3).

Figure 3. Result of flownet tracking.

After this rotation was discovered, a "point tracking" function was used to monitor the material flow on the cross-section of the stock. A total of six points were marked on a radial line through the stock axis, starting on the axis and spaced by 3 mm in direction to the surface (Figure 4. a,b). Rolling caused relative movement of these points (Figure 4. c,d). On longitudinal sections, the points which had been closer to the surface shifted in the rolling direction. On transverse sections, the point on the surface travelled relative to the others in the direction indicated on the flownet tracking plot in Figure 3.

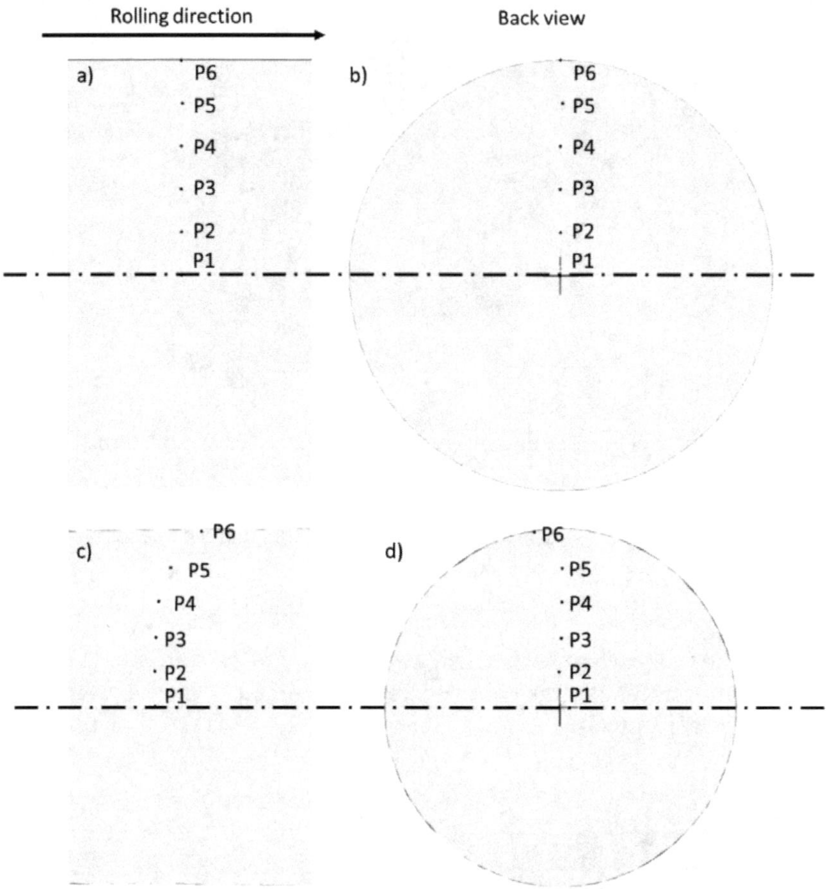

Figure 4. Point tracking analysis: a) longitudinal section before rolling, b) cross-section before rolling, c) longitudinal section after rolling, d) cross-section after rolling.

3.2 Effective stress

Stresses on the contact surface between the rolls and the stock were examined. As expected, the values were higher than in the core of the stock (Table 2). In the carbon steels, the highest stress levels were between 165 MPa in the C60 grade and 210 MPa in C15. In AISI 321, the maximum stress levels increased with strain rate.

Table 2. The highest effective stress in the bar surface during rolling.

Material	C15	C45	C60	AISI 321 - 50 rpm	AISI 321	AISI 321 - 200 rpm
Effective stress [MPa]	~210	~185	~165	~220	~255	~320

Furthermore, a radial profile of stress in the rolled stock was examined (Figure 5). Stress increased with the distance from the axis in all the materials and for all rolling speeds. Stress values were lower in steels with less carbon. In AISI 321 grade rolled at 50 rpm, the stress in the core was lower than in the carbon steels, although the stress level near its surface was higher than in the carbon grades. There is a greater difference between the minimum and maximum stress in this stainless steel.

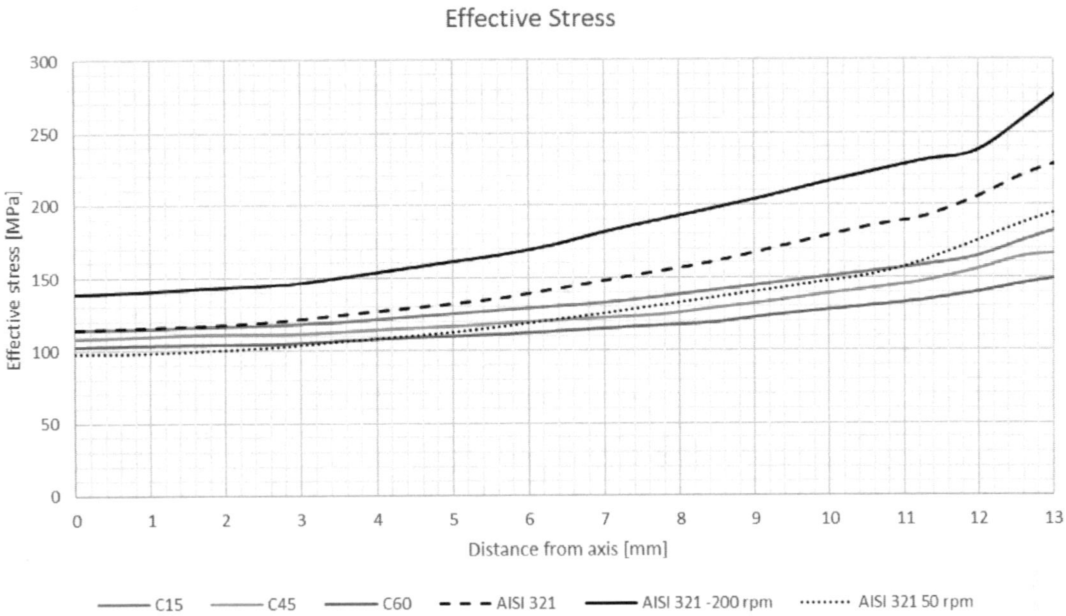

Figure 5. The radial profile of effective stress in the rolling gap between the bar axis and surface.

3.3 *Effective strain*

Finally, effective strain and its distribution on the cross-section through the rolled stock was examined before entry into and after exit from the rolling gap. The effective strain field on several cross-sections through the rolled portion of the stock indicates gradual changes in its roundness and increasing strain (Figure 6.). Fig. 6 illustrates notable differences between effective strain in AISI 321 and C60, namely in the surface layers where the levels were higher in C60, although the core strain was similar in both grades.

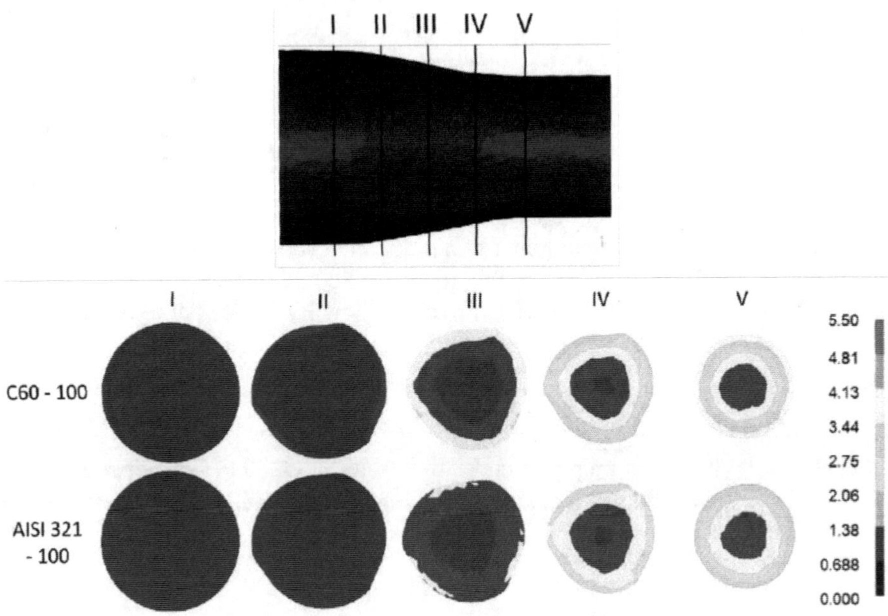

Figure 6. Effective strain during rolling

Effective strain levels near the axis of the stock are almost equal in all the cases: 0.75 (Figure 7.). They increase in a similar trend with the distance from the axis for all the process sequences. Differences begin to appear about 5 mm below the surface of the stock. Between this location and the surface, carbon steels reach higher effective strains than AISI 321.

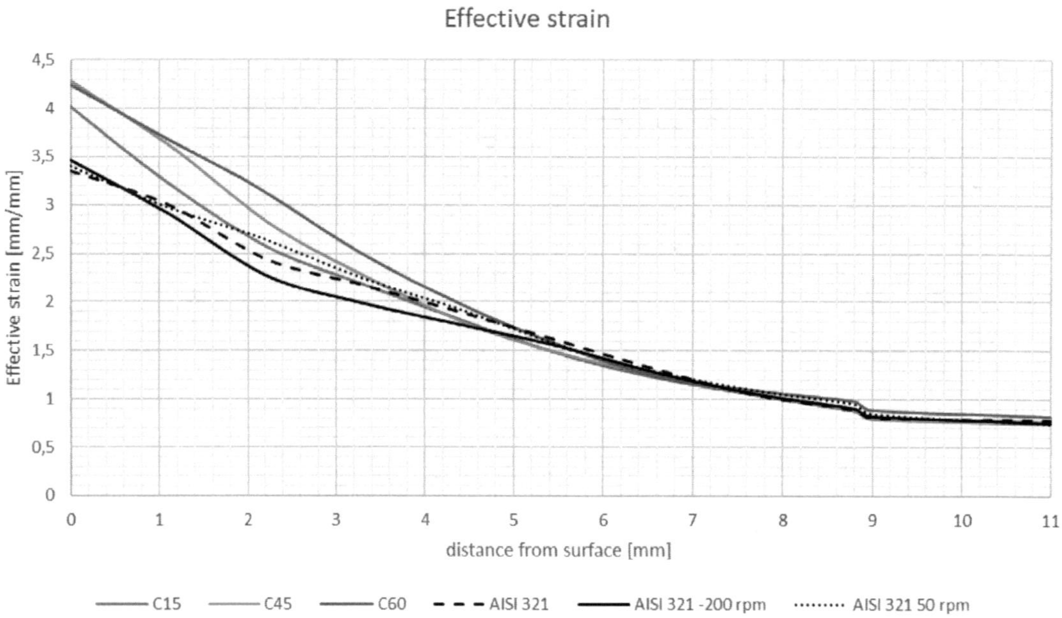

Figure 7. Effective strain distribution after rolling.

4 Conclusion

Numerical modelling of a round bar rolling process in HDQT-R 30-12 machine developed using DEFORM 2D/3D software revealed that the three rolls cause the stock material to flow in the rolling direction, i.e. forward, as well as in the rotation direction of the stock. According to stress analysis, higher stresses are induced in steels with less carbon than in those with higher carbon contents. The reason is that steels with less carbon have higher flow stress values at these rolling temperatures. In the stainless AISI 321 grade, stress increased with strain rate. Effective strains on the cross-section near the axis of bars are nearly identical for all the materials and increase with distance from the axis. However, they increase differently in each material. In carbon steels, the effective strain near the surface is larger than in the stainless grade, regardless of the rolling speed. Effective strain accumulates in the surface of the rolled bar which can be seen on both longitudinal and transverse sections. It suggests that thanks to low effective strains along the axis of the bar, this equipment could be used for rolling tubes without the use of a mandrel and risking their collapse. This hypothesis will be tested and the outcomes published if relevant.

Acknowledgement

The present contribution has been prepared under the project LO1502 'Development of the Regional Technological Institute' under the auspices of the National Sustainability Programme I of the Ministry of Education of the Czech Republic aimed to support research, experimental development, and innovation.

References

[1] Fan J, Tang C, Tsui C, Chan L, Lee T 2006 3D finite element simulation of deep drawing with

damage development *J. International Journal of Machine Tools & Manufature* (Elsevier) **46** p 1035-1044

[2] Gouveia B, Rodrigues J, Martins P 1995 Fracture predicting in bulk metal forming *J. International Journal of Mechanical Science* (Great Britain: Elsevier Science Ltd) **4** p 361-372

[3] Rząsowska-Przała A, Waszkielewicz W, Gałkin A 2005 Otrzymywanie Prętów za pomocą walcowania na gorąco w walcach skośnych w porównaniu z innymi sposobami ich wytwarzania *J. Zagadnienia techniczno-ekonomiczne* (Krakow, Akademia Górniczo-Hutnicza Im. Stanisława Staszica) **50** p 169 – 173

[4] Nakasuji K, Kuroda K, Hayashi C 1996 Reduce Rolling Characteristics of Hollow Piece by Rotary Rolling Mill *J. ISIJ International* (Japan, The Iron and Steel Institute of Japan) **36** p 572 - 578

[5] Stefanik A, Morel A, Mróz S, Szota P 2015 Theoretical and Experimental Analysis of Aluminium Bars Rolling Process in Three.High Skew Rolling Mill *J. Archives of Metallurgy and Materials* (Poland, Polish Acamecy of Sciences) **60** p 809 – 813

[6] GMT mbH 2015 Provozní návod zařízení HDQT – R 30 – 12 pro termomechanické zpracování tyčové oceli (Berlin: Gesellshaft für metallurgishe technologie und softwareentwicklung)

[7] Pekovic M, Vrtacek J, Janda T, Volkmannová J 2020 Induction hardening of steels with use of the device for incremental forming of round bars HDQT-R 30-12 *J. IOP Conference Series Materials Science and Engineering* (IOP Publishing Ltd) **723**

Scanning electron microscopy as a useful tool for the analysis of non-conductive materials

Iveta Tichá[1] and Ludmila Kučerová[1]

[1]Laboratory of Metallography, Regional Technological Institute, University of West Bohemia in Pilsen, Univerzitní 22, 306 14, Plzeň, Czech Republic

E-mail: ichenati@rti.zcu.cz, skal@rti.zcu.cz

Abstract. Scanning electron microscopy (SEM) in the analysis of non-conductive samples became one of the most important methods for the investigation of material properties. In this work, we used SEM microstructure analysis for the investigation of the origin of cracks in granite composites and also, we tested the porosity inside the regenerated carbon biowaste, potentially used as a clean source of carbon for the future applications in materials production. Additionally, the morphology and the chemical composition of small particles used for the moulding processes of plastics were also tested. The importance of the microstructure investigation was supported by Energy-Dispersive X-ray Spectroscopy (EDS) often used for the chemical composition evaluation of these non-conductive materials.

Key words: Scanning electron microscopy, Energy-Dispersive X-ray Spectroscopy, Non-conductive materials, Fractography, Carbon porosity

1 Introduction

Scanning electron microscopy (SEM) represents a powerful technique for the microstructure and morphology analysis of a wide range of materials [1,2]. Considering the advanced material analysis techniques nowadays, SEM method can be used not only for the conductive materials but also for the non-conductive samples and even for the biological tissues [3,4]. However, this application strongly depends on the technique of SEM and also on the preparation of tested materials [5]. On the other hand, the investigation of such samples contributes to the problem solutions in the material industry and other applications [1]. For instance, SEM microstructure analysis can easily reveal the origin of the cracks in materials or the porosity of small particles. In combination with Energy-Dispersive X-ray Spectroscopy (EDS) analysis, this creates the powerful tool for detail analysis of various materials [6]. Thus, in this work, we focused on three different types of non-conductive samples which were analysed by SEM and EDS. The chosen samples were industry-oriented, for instance the origin of cracks in granite composites or the porosity measurements in the regenerated biowaste materials and the analysis of plastic granules.

2 Experimental Section

2.1 Microstructural characterisation

Detailed metallographic analysis was performed by scanning electron microscope Zeiss EVO 25 with a LaB_6 cathode. SEM images were performed under the lower pressure in system (25 Pa) with 15 kV voltage, however EDS analyses were performed under 20kV voltage and 7 nA I-probe, both with the easy-VP aperture in the VP (variable pressure) mode. This VP mode works at higher pressures (10-3000 Pa) in the instrument chamber and thus differs from the conventional SEM operation under lower pressure which usually requires coatings of the non-conductive samples [7]. Both SE and BSE detectors were used for the analysis. The metallographic cross-sections of the chosen samples were prepared conventionally with the mounting and polishing. However, the specimens were not etched. Chemical compositions of the samples were locally investigated by EDS (Oxford Instruments).

2.2 Materials

Various non-conductive materials were tested in our laboratory, including the analysis of crack initiation and defects, and also new material testing which is the most typical analysis of metallographic laboratory. Nevertheless, in this work we focused only on three, industry-oriented types of materials. More specifically, the first materials for analysis were granites composites, mainly used in house-hold equipment. The granite composites were tested in the form of the angular and round house-hold sinks which cracked during the service and were further cut into small samples for the analysis (Figure 1).

a)
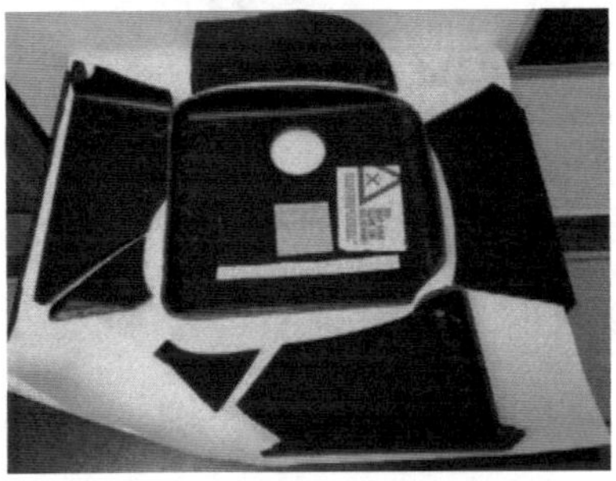

b)
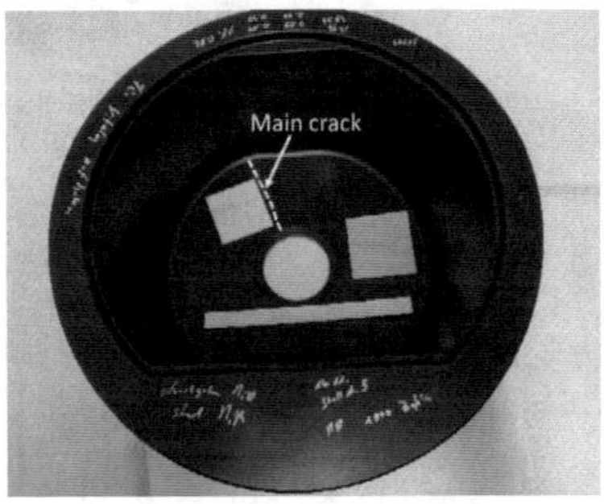

Figure 1. Overview of granite composites (a-angular and b-round sinks) used for the analysis.

Subsequently, the porosity of the carbon biowaste was tested in the chosen samples, specifically in the regenerated coffee mixtures (Figure 2a), processed with different inorganic compounds for the activation (such as Na_2CO_3 or H_3PO_4). Finally, as-delivered plastic TECHNYL® cylinder-like shape particles were tested and the origin of the small fibres on the surface of these particles was investigated (Figure 2b).

a)
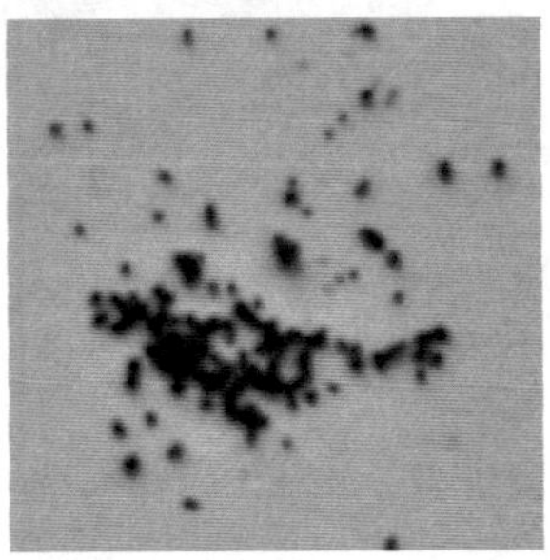

b)

Figure 2. Overview of the regenerated coffee grounds (a) and TECHNYL® granules (b).

3 Results and Discussion

2.1 SEM analysis of granite composites cracks

Granite composites became a good alternative for the stainless-steel materials, especially in a house-hold equipment such as sinks [8]. Nevertheless, the widespread usage of these composites is hindered with the insufficient identification of the defects and morphology analyses [9]. Thus, it is necessary to investigate the cracks and other defects and their origins should be identified to understand and avoid the mistakes during the production processes.

Due to high fragmentation of the angular sink (Figure 3a), it was quite hard to determine the area of the fracture origin. Nevertheless, we chose the part from the bottom edge which is the probable area of the fraction origin (or one of the areas of fracture origin) and also its surface was appropriate for investigation by SEM. On the other hand, the round sink (Figure 3b) was investigated at the bottom part, other parts were intact. At the bottom part, all the cracks have radial orientation towards the centre hole of the sink.

a) 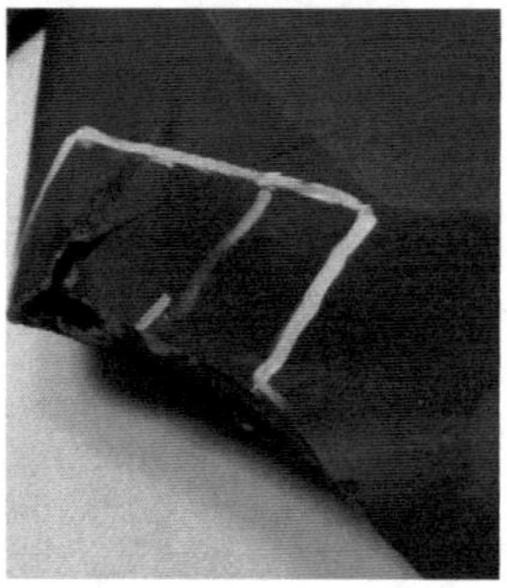 b) 

Figure 3. Presumed initiation areas of cracks in the granite-composite sinks (a-angular, b-round).

Initially, we investigated the metallographic cross-sections (in the plane perpendicular to the fracture surface) of the angular sink to understand the microstructure and morphology of the material (Figure 4). This microstructure contains the mixture of relatively large particles homogenously distributed in the matrix.

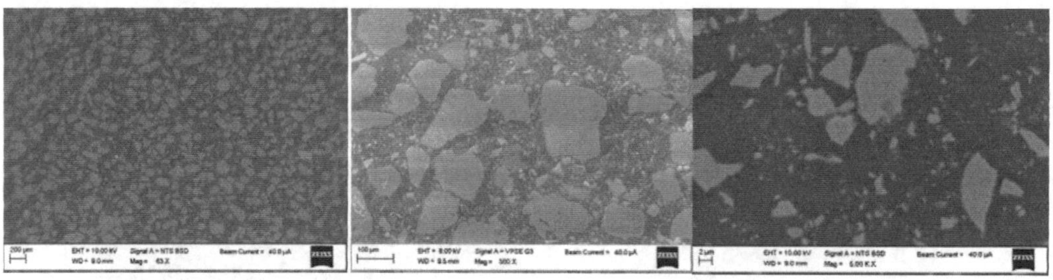

Figure 4. SEM images of metallographic section of the angular granite-composite sink.

The EDS analysis revealed that the Al-fine particles were distributed through the large, Si-based particles (Figure 5). However, the EDS analysis can detect only the elements with local content more than 0.1 weight % [10].

IOP Conf. Series: Materials Science and Engineering 1161 (2021) 012005 doi:10.1088/1757-899X/1161/1/012005

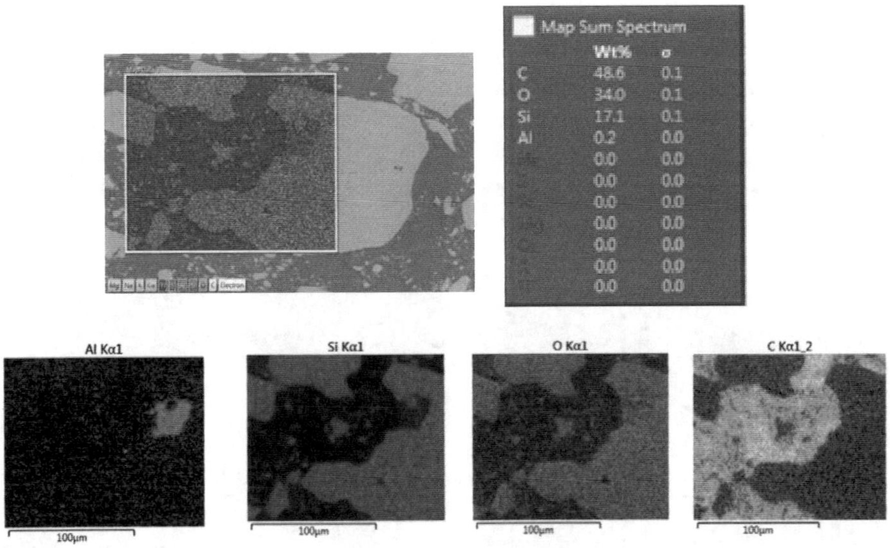

Figure 5. EDS of metallographic section of angular sink, mapping, average chemical composition.

Subsequently, the fracture initiation area was found in the bottom edge of the angular sink, from where a large secondary crack also propagated through the bottom of the sink (red circle on the left, Figure 6). The layer of the melted material was observed at fracture plane, just below the surface of the sink (red arrow in the middle and on the right, Figure 6). This unique feature was not found in any other areas of the fracture surface and it was also not observed at metallographic section. Overall, this material defect could be assumed to cause fracture initiation. The chemical composition was also investigated by point and map EDS analysis.

Figure 6. The crack initiation area of the angular sink, SEM and EDS analysis.

On the other hand, the character of damage in the round sink was completely different. As discussed previously, several cracks were observed only at the bottom of the otherwise intact sink, covering slightly more than one fourth of the bottom area. Crack initiation area showed brittle defragmentation of Si-based particles which resulted in the crack initiation at a particle-matrix interface (Figure 7). Crack propagation was further supported by thermo-mechanical fatigue. In comparison to the angular sink, the round sink showed more areas with locally increased Ti and occasionally also Fe contents. However, the average Ti and Fe contents were still below 0.1 weight % and these elements were therefore not detected by plane EDS analysis.

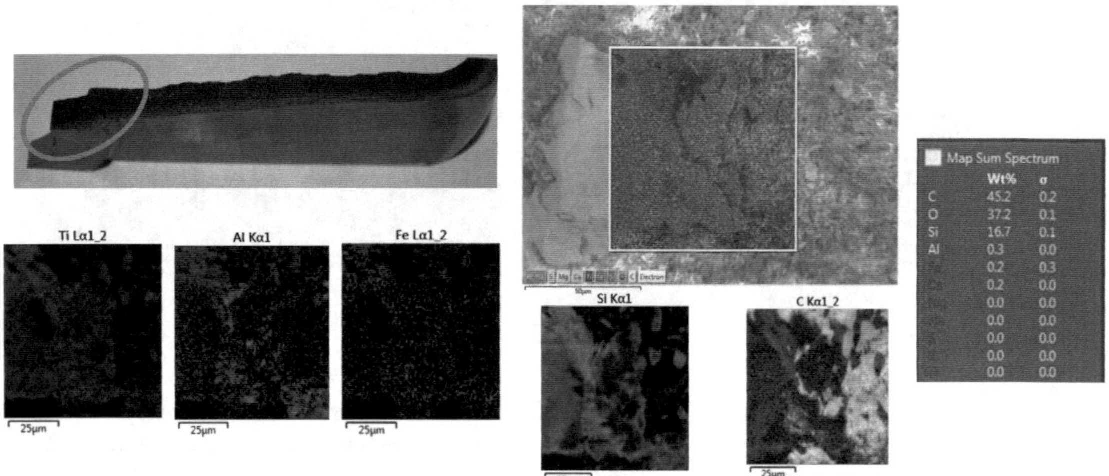

Figure 7. The crack initiation of the round sink and EDS chemical analysis.

2.2 SEM analysis of a porosity of regenerated coffee grounds

The importance of regenerated carbon biowaste, such as that from the coffee or tea grounds has increased recently, as the renewable source of porous carbon plays an important role in a wide range of applications such as bioplastics, biofuels and other productions [11], [12]. However, the activation of such regenerated particles of carbon deals with the problems in cleaning and the size of the pores. Thus, these pores in the grounds after different activation processes (in acidic or base conditions) were evaluated and compared (Figure 8). SEM images showed that the basic conditions (activation with Na_2CO_3 in Figure 8a) led to creation of larger pores of better quality than the activation under acidic conditions which resulted in smaller and poor-quality pores in coffee grounds (activation with H_3PO_4 in Figure 8b). The parameters of both activations are included in [13].

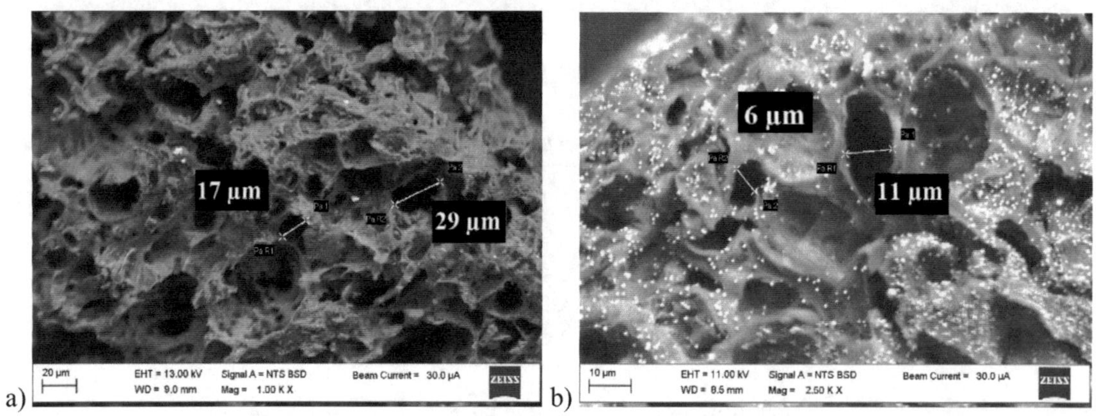

Figure 8. SEM analysis of coffee grounds and the evaluation of the porosity (a-Na_2CO_3, b-H_3PO_4).

2.3 *SEM analysis of plastic granules and fibres*

Homogeneity of plastic particles presents the key property for the nonproblematic production process of the various plastic items [14]. However, sometimes the storage of such granules introduces the problem of secondary particles or fibres which is necessary to investigate [15]. Thus, the analysis of the origin of the heterogeneous objects created during the manipulation with the plastic granules revealed that the small fibres were confirmed in the whole volume of storage package. Specifically, the fibres were observed on the edge of the granules (Figure 9). Furthermore, EDS analysis revealed that these fibres had the same chemical composition as the primary granules.

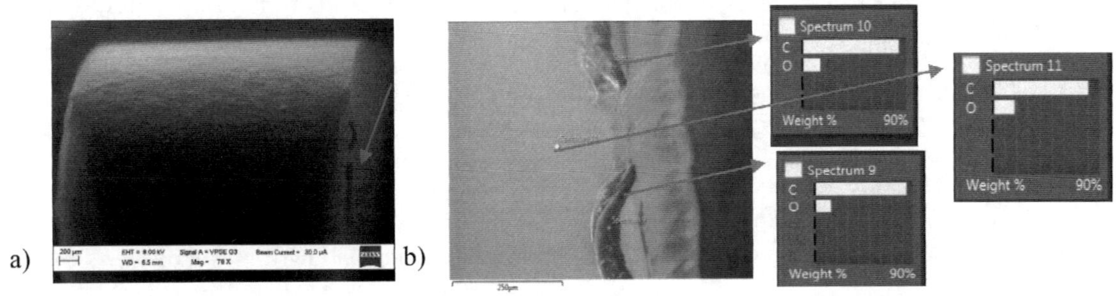

Figure 9. SEM and EDS analyses of the plastic granules and fibers (a-overall view of the granule, b-detail of the part with the fiber).

3 Conclusion

Without any doubt, SEM and EDS techniques represent the fast and precise investigation of non-conductive materials. They are used not only for the analysis of the origins of cracks but also to determine the microstructure and chemical analysis of plastic granules or biowaste-carbon particles, etc. Specifically, in this work, the results obtained from SEM images clearly confirmed that the cracks in the tested granite composites have different origins such as the brittle defragmentation Si-based particles on one side or the thermo-mechanical fatigue on the other side. Further, SEM images also revealed that the porosity of regenerated coffee grounds strongly depends on the chemical conditions (acidic or basic environments) of the activation process. Finally, SEM analysis of plastic particles and fibres on theses particle surfaces confirmed the same chemical conditions for both parts.

Acknowledgements

The present contribution has been prepared under project LO1502 'Development of the Regional Technological Institute' under the auspices of the National Sustainability Programme I of the Ministry of Education of the Czech Republic aimed to support research, experimental development and innovation.

References

[1] Reimer L 2000 Scanning Electron Microscopy: Physics of Image Formation and Microanalysis, Second Edition, Meas. Sci. Technol. 11 1826–1826.

[2] Goldstein J I Newbury D E Michael J R Ritchie N W M Scott J H J Joy D C Scanning Electron Microscopy and X-Ray Microanalysis, Springer New York, New York, NY, 2018.

[3] Cao C Gao R Shang H Peng T Characterization of non-conductive materials using field emission scanning electron microscopy, in: L. Yu (Ed.), Xia'men, China, 2016: p. 990319.

[4] Wolff A Zhou Y Lin J Peng Y Y Ramshaw J A M Xiao Y FIB/SEM Processing of Biological Samples, Microsc Microanal. 24 (2018) 822–823.

[5] Hall J L ed., 1991 Electron microscopy of plant cells, Acad. Press, London.

[6] Shim W J Hong S H Eo S E Identification methods in microplastic analysis: a review, Anal. Methods. 9 (2017) 1384–1391.

IOP Conf. Series: Materials Science and Engineering 1161 (2021) 012005 doi:10.1088/1757-899X/1161/1/012005

[7] Thiel B Variable Pressure Scanning Electron Microscopy, in: P.W. Hawkes, J.C.H. Spence (Eds.), Springer Handbook of Microscopy, Springer International Publishing, Cham, 2019: pp. 319–344.
[8] Ramakrishna H V Priya S P Rai S K Varadarajulu A Tensile, Flexural Properties of Unsaturated Polyester/Granite Powder and Unsaturated Polyester/Fly Ash Composites, Journal of Reinforced Plastics and Composites. 24 (2005) 1279–1287.
[9] Piratelli-Filho A Levy-Neto F Behavior of granite-epoxy composite beams subjected to mechanical vibrations, Mat. Res. 13 (2010) 497–503.
[10] Nasrazadani S Hassani S Modern analytical techniques in failure analysis of aerospace, chemical, and oil and gas industries, in: Handbook of Materials Failure Analysis with Case Studies from the Oil and Gas Industry, Elsevier, 2016: pp. 39–54.
[11] Xu C Nasrollahzadeh M Selva M Issaabadi Z Luque R Waste-to-wealth: biowaste valorization into valuable bio(nano)materials, Chem. Soc. Rev. 48 (2019) 4791–4822.
[12] Kemp K C Baek S B Lee G W Meyyappan M Kim K S Activated carbon derived from waste coffee grounds for stable methane storage, Nanotechnology. 26 (2015) 385602.
[13] Forejtová J Diploma thesis, University of West Bohemia in Pilsen, 2020.
[14] Hassan A Hornsby P R Folkes M J Structure–property relationship of injection-molded carbon fibre-reinforced polyamide 6,6 composites: the effect of compounding routes, Polymer Testing. 22 (2003) 185–189.
[15] Guise O Strom C Preschilla N STEM-in-SEM method for morphology analysis of polymer systems, Polymer. 52 (2011) 1278–1285.

IOP Conf. Series: Materials Science and Engineering 1161 (2021) 012006 doi:10.1088/1757-899X/1161/1/012006

Design of thermomechanical treatment of high strength steels

J Vrtáček[1], M Peković[2], H Jirková[3], D Bublíková[4] and L Kučerová[5]

UWB in Pilsen, Regional Technological Institute, Univerzitní 22, 301 00 Pilsen, Czech Republic

[1]vrtacekj@rti.zcu.cz,[2]pekovicm@rti.zcu.cz, [3]hstankov@rti.zcu.cz, [4]natasha@rti.zcu.cz, [5]skal@rti.zcu.cz

Abstract. Reducing the level of COx and NOx emissions from combustion engines is an important goal in the automotive industry. One way to reduce the emissions is to use lightweight parts in the car construction. High-strength steels are one of the successful choices for body parts. Their strengthening mechanisms provide them with better mechanical properties, which translates into thinner sheet metal blanks and ultimately reduced weight of parts. The main strengthening mechanisms include TRIP (TRansformed Induced Plasticity) and TWIP (TWinning induced plasticity). This paper outlines processing of high-strength steels, from forging to round bars and heat treatment to determination of phase transformations and temperature profiles for thermomechanical processing in an innovative tool. The experimental materials contained manganese, silicon and different aluminium levels. Their compositions were selected with a goal to observe the properties of advanced multiphase high strength steels during processing.

1 Introduction

Modern high-strength steels have an important role in enhancing passenger safety. The purpose of vehicle parts made from these materials is to absorb the largest possible portion of crash energy while minimizing plastic deformation in the deformation-absorbing parts of the vehicle body or frame. These advanced high-strength steels rely on a combination of several strengthening mechanisms, including the TRIP (TRansformation Induced Plasticity) and TWIP (TWinning Induced Plasticity) effects [1, 2]. These mechanisms require retained austenite to be present and rely on lattice transformation induced by plastic deformation [1,10]. TRIP effect consists in the transformation of retained austenite (RA) to martensite, whereas the TWIP mechanism produces twins in austenite. Once the mechanical properties of the material are improved, the designers can reduce the thickness of sheet blanks. Along with thickness, the final weight of the body or frame of the vehicle as well as the deformation-absorbing parts can be reduced, which leads to less emissions. Recently, however, the sheets already have thicknesses which are close to the acceptable limit (0.8–1.2 mm) which impacts the part's performance. Where further reduction is required, it is achieved by alloying the material with light metals, such as aluminium [3]. Aluminium reduces the density of the matrix, has favourable effects on grain size obtained by heat treatment and can substitute silicon to some extent [4]. The past decade has seen the development of high-strength multiphase steels of a new generation which combine the benefits found in the previous generations. They contain manganese and aluminium. It is the manganese-aluminium ratio plays a key role in obtaining the desired mechanical properties and the forming behaviour. This study therefore focuses on two advanced high-strength steels with different aluminium levels in an attempt to characterize the differences between their behaviour during forming and heat treatment.

2 Materials and methods

2.1 Experimental materials

Two steel grades were studied: C3Mn2Al and C3Mn1.4Al. The C3Mn1.4Al grade contained 0.22% C, 3.06% Mn, 0.59% Si, 1.45% Al, 0.008% P and 0.003 S. For the sake of comparison, the other material had a higher level of aluminium, which was 2%. Manganese [12] is added as an austenite-forming element which does not tend to form carbides, i.e. particles which are undesirable for the final thermomechanical treatment to make the finished product [6]. Manganese also improves elasticity of the material, as it dissolves in ferrite. Its atoms cause distortion of the lattice and prevent dislocation movement [5, 11]. The role of aluminium is to act as a ferrite-former and to provide deoxidation [5,6].

In order to collect extensive data, the measurement of mechanical properties (tensile and hardness testing) was complemented with diffraction phase analysis by which the fraction of retained austenite was determined. The measurement was carried out using the AXS Bruker D8 Discover automatic powder diffractometer with a HI-STAR position-sensitive area detector and a cobalt X-ray source ($\lambda_{K\alpha}$ = 0.1790307 nm). Measurements were taken at the centres of metallographic sections at diffraction angles in the interval of 25 ÷ 105°2ϑ.

2.2 Determination of phase transformation temperatures

Prior to hot rolling were the top and bottom of the ingots removed The ingots were placed into a pre-heated furnace at 1050 °C for 2 hours. Then they were transferred from the furnace to the rolling mill. The rolling sequence consisted of nine successive reductions with no reheating operation. The initial diameter was 97 mm and the final thickness of the strip was 13 mm. The temperature at the entry was about 1035 °C and temperature did not dropped below 950 °C during all reductions. With each reduction, the rolling force increased: from the initial 680 kN to 2000 kN in the last reduction. No excessive buildup of force, irregular thickness or waviness were experienced in the rolling process. Immediately after finish rolling to 13 mm, the rolled part was cut into three identical segments 800 mm in length and reheated in a furnace for 1 hour.

To design the heat treatment sequences for this experiment, one had to determine phase transformation temperatures for both materials. They were calculated from their chemical composition using JMatPro software [7]. In an alternative procedure, Andrews equations for calculating phase transformation temperatures were employed (1 through 3) [8]. There was a good agreement between the calculated values and the data from the dilatometer (table 1). Yet, there were differences in A_{c3} temperature, up to 200°C. It was because the calculation did not account for all alloying elements in the material. The effect of aluminium as a ferrite-former was manifested in the A_{c3} temperature which was higher in C3Mn2Al.

$$Ms(°C) = 539 - 423C - 30{,}4Mn - 17{,}7Ni - 12{,}1Cr - 7{,}5Mo \quad (1)$$

$$A_{c3}(°C) = 910 - \sqrt{203C} + 44{,}7Si - 15{,}2Ni + 31{,}5Mo + 104{,}4V + 13{,}1W \quad (2)$$

$$A_{c1}(°C) = 723 - 10{,}7Mn - 16{,}6Ni + 29{,}1Si + 16{,}9Cr + 290As + 6{,}38W \quad (3)$$

Table 1. Phase transformation temperatures obtained from equations, JMatPro software and dilatometric measurement.

	C3Mn1.4Al			C3Mn2Al		
	JMatPro	Andrews	Dilatometer	JMatPro	Andrews	Dilatometer
M_s [°C]	351	349	342	355	346	328
A_{c3} [°C]	815	841	961	981	841	1025
A_{c1} [°C]	709	691	728	714	691	739

2.3 Heat treatment design

After breaking down and forging the C3Mn1.4Al and C3Mn2Al ingots into the required feedstock, an appropriate austenitizing temperature had to be found for the heat treatment. Specimens of the steels were brought to different temperatures between 900 and 1200°C set at 50°C intervals. For each temperature, the holding time in the furnace was 30 minutes.

Austempering trials were the next step. The specimens were brought to 1000°C and held for 20 minutes in a furnace, after which they were quenched in a salt bath at 400°C. Once they reached 400°C, they were removed from the bath. The temperature was measured with a K-type thermocouple which was inserted in a hole in the centre of the specimen. After removal from the bath, the specimens were held in a furnace at 400, 425 or 450°C. Two different holding times were used: 15 and 60 minutes. For both steels, the annealing temperatures were in the bainitic transformation region and above the Ms. In the salt bath, the cooling rate was 35°C/s, as determined from the thermocouple data.

2 Results and discussion

3.1 Determination of suitable austenitizing temperature

The resulting microstructures were observed using light microscopy. In both materials, austenitizing at the lowest temperature, 900°C, led to a martensitic-bainitic microstructure with a small amount of ferrite (figure 1 a,b). C3Mn1.4Al had a hardness of 513 HV10. In C3Mn2Al, it was 476 HV10. Austenitizing at the next higher temperature, 950°C, led to different microstructures. Martensite was the prevalent phase (figure 1 c,d), which led to a higher hardness of 535 HV10 in C3Mn1.4Al and 505 HV10 in the C3Mn2Al grade. After austenitizing at still higher temperatures, the martensitic structure became coarser. Nevertheless, its hardness was not any higher (figure 2). The difference in hardness of the experimental materials is due to the different aluminium levels, since this element affects the transformation behaviour and mechanical properties.

IOP Conf. Series: Materials Science and Engineering 1161 (2021) 012006 doi:10.1088/1757-899X/1161/1/012006

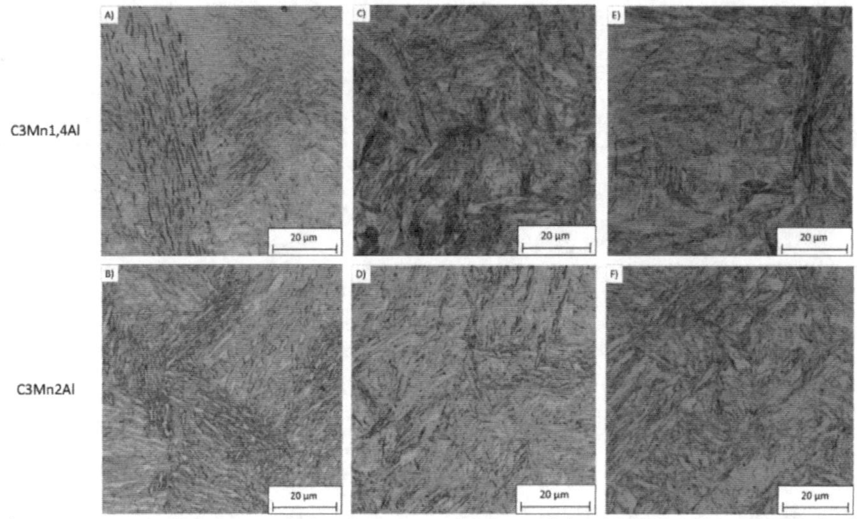

Figure 1. Microstructures of the materials after austenitizing trials: 900°C (a,b), 950°C (c,d) and 1200°C (e,f).

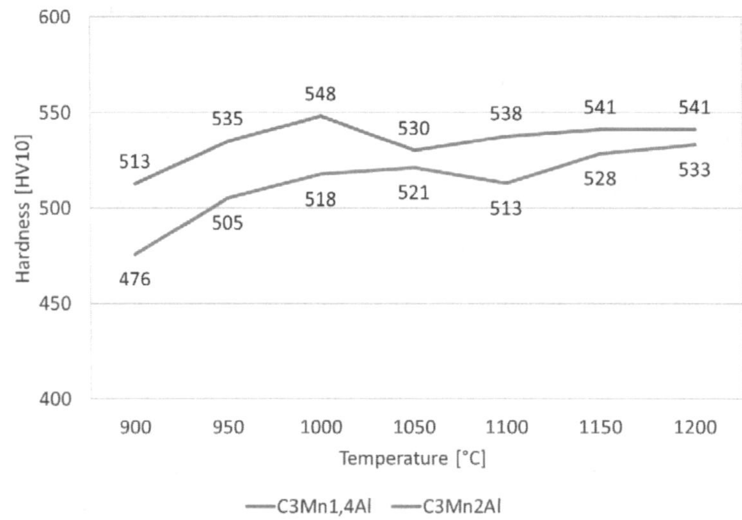

Figure 2. Hardness of the experimental materials C3Mn1.4Al and C3Mn2Al after austenitizing trials.

3.2 Austempering

The purpose of the austempering trials was to determine the effect of the aluminium level, temperature and holding time at a temperature in the bainitic transformation region.

Austempering at 400°C produced a bainitic-martensitic microstructure in C3Mn1.4Al (figure 3) with an ultimate strength of 1220 MPa, elongation of 25% (table 2) and a hardness of 387 HV10 (figure 6). The retained austenite volume fraction was 15 % (figure 7). The microstructure of C3Mn2Al was similar to the preceding one: bainite-martensite, with some bainitic ferrite and visible prior austenite grain boundaries. Its strength was 1390 MPa, elongation reached 17% (table 2), hardness was 424 HV10 (figure 6) and the volume fraction of retained austenite was 9% (figure 7). When the specimens were held for 60 minutes instead of 15 minutes, the final bainite fraction was higher. This led to a lower ultimate strength of 1220 MPa (whereas at 15 minutes, it was 1260 MPa in the C3Mn1.4Al grade). The resulting hardness of 376 HV10 was lower than in the previous specimen (387 HV10). However, elongation A_{5mm} [14] was higher: 22%, as opposed to 17%. The volume fraction of retained austenite decreased to 16%. With the other material, C3Mn2Al, the trend was the same.

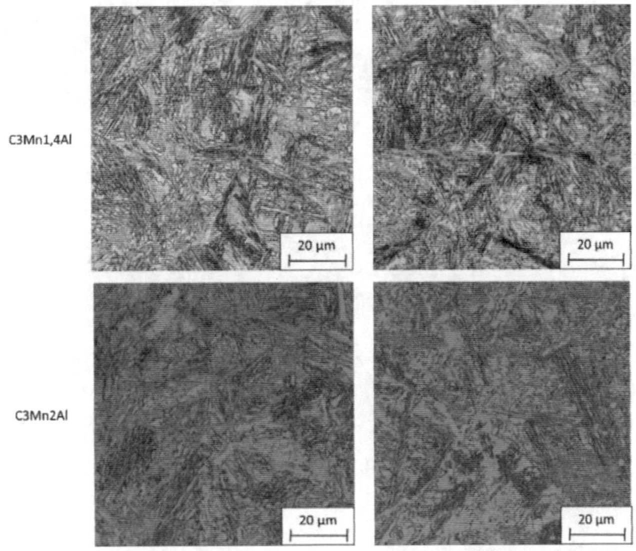

Figure 3. Microstructures of the materials after austempering at 400°C with 15 minute (left) and 60 minute (right) holding times.

Table 2. Mechanical properties of the materials after austempering at 400°C.

Experimental material	Treatment route	Offset yield strength ($R_{p0.2}$) [MPa]	$R_{p0.2}$ standard deviation	UTS (R_m) [MPa]	R_m standard deviation	A_{5mm} [%]	A_{5mm} standard deviation
C3Mn1.4Al	400°C - 15 min.	683	7.1	1257	7	19.3	0.008
	400°C - 60 min.	673	6.8	1223	2	25.3	3.1
C3Mn2Al	400 °C - 15 min.	824	19	1386	8	16.6	1.0
	400°C - 60 min.	631	4	1220	5	21.1	0.7

After austempering at the higher temperature of 425°C, the C3Mn1.4Al grade developed a bainitic-martensitic microstructure with a larger amount of martensite and less bainitic ferrite (figure 4) than after the previous treatment at 400°C. The strength after the sequence with a holding time of 15 minutes was higher: 1380 MPa. Elongation reached 15% (table 3) and hardness was 428 HV10 (figure 6). The amount of retained austenite was 8% (figure 7). After the 60-minute route, the ultimate strength was higher, 1360 MPa, and elongation and hardness were 15% and 400 HV10, respectively, whereas the amount of RA was less: 8%. The C3Mn2Al grade had a bainitic-martensitic microstructure with some bainitic ferrite and 9% retained austenite. The ultimate strength was between 1360 and 1375 MPa, the elongation was 15 % (table 3) and hardness reached 408 HV (figure 6).

After intercritical annealing at 425°C, mechanical properties were higher but elongation remained unchanged in both materials. The retained austenite fraction decreased to 8–11% (figure 7). The final microstructure contained bainite and martensite at a well-balanced ratio (figure 4). After this treatment, the material attained the properties required in final products for safety-relevant parts of the car body.

Figure 4. Microstructures of the materials after austempering at 425 °C with 15 minute (left) and 60 minute (right) holding times.

Table 3. Mechanical properties of the materials after holding at 425°C.

Experimental material	Treatment route	Offset yield strength ($R_{p0.2}$) [MPa]	$R_{p0.2}$ standard deviation	UTS (R_m) [MPa]	R_m standard deviation	A_{5mm} [%]	A_{5mm} standard deviation
C3Mn1.4Al	425°C - 15 min.	893	69	1379	14	15.3	0.04
	425°C - 60 min.	727	43	1238	100	15.0	2.3
C3Mn2Al	425°C - 15 min.	765	9	1364	4	14.9	2.0
	425°C - 60 min.	730	1	1361	6	13.5	1.0

After holding at 450°C, the amount of martensite in the martensitic-bainitic structure C3Mn1.4Al was higher still (figure 5). The ultimate strength increased to 1400 MPa, elongation was 14–18% (table 4) and hardness was close to 450 HV10. The retained austenite fraction was 6% after the route with the shorter holding time and 9% after the 60-minute hold. In C3Mn2Al, the microstructure consisted of martensite and bainite as well, with a higher proportion of martensite and some free ferrite (figure 5). The ultimate tensile strength and hardness were 1400 MPa and 431 HV10, respectively, after holding for 15 minutes. After the 60-minute hold, these values were 1424 MPa and 438 HV10. Elongation was approximately 14%.

Treatments at higher holding temperatures led to higher mechanical properties and less retained austenite.

IOP Conf. Series: Materials Science and Engineering 1161 (2021) 012006 doi:10.1088/1757-899X/1161/1/012006

Figure 5. Microstructures of the materials after austempering at 450°C with 15-minute (left) and 60-minute (right) holding times.

Table 4. Mechanical properties of the materials after holding at 450°C.

Experimental material	Treatment route	Offset yield strength ($R_{p0.2}$) [MPa]	$R_{p0.2}$ standard deviation	UTS (R_m) [MPa]	R_m standard deviation	A_{5mm} [%]	A_{5mm} standard deviation
C3Mn1.4Al	450°C - 15 min.	860	10	1370	13	14.6	0.9
	450°C - 60 min.	888	13	1394	1	13.7	0.4
C3Mn2Al	450°C - 15 min.	811	5	1402	8	13.8	0.1
	450°C - 60 min.	851	11	1424	2	13.2	0.5

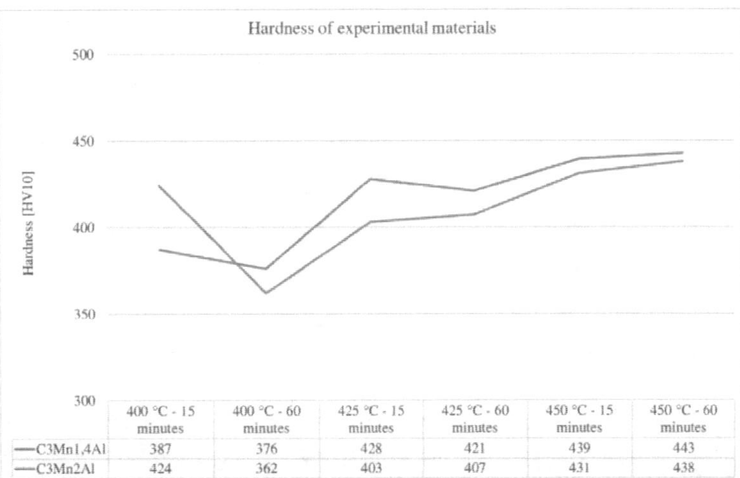

Figure 6. Effects of heat treatment on hardness in C3Mn1.4Al and C3Mn2Al.

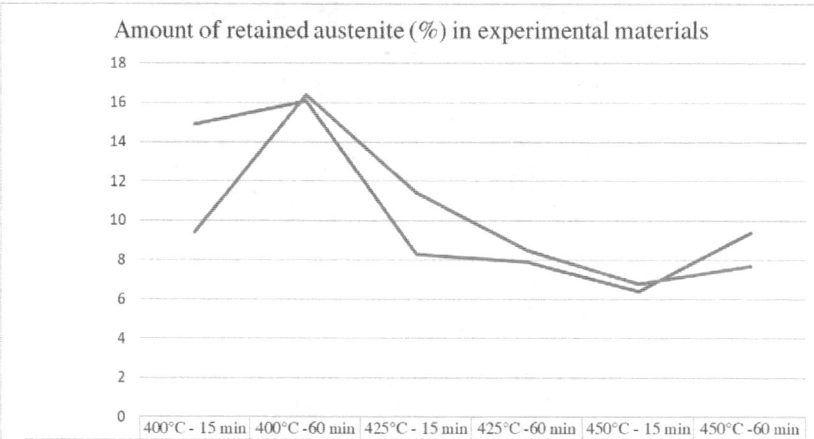

	400°C - 15 min	400°C -60 min	425°C - 15 min	425°C -60 min	450°C - 15 min	450°C -60 min
—C3Mn1,4Al	14,9	16,1	8,3	7,9	6,4	9,4
—C3Mn2Al	9,4	16,4	11,4	8,5	6,8	7,7

Figure 7. The proportions of retained austenite after heat treatment.

4 Conclusion

Intercritical annealing routes were designed for C3Mn1.4Al and C3Mn2Al grades. Breaking down and forging of ingots into round bars was trouble-free for C3Mn2Al. However, the other material suffered irreparable damage during heating to the forging temperature of 1150°C. Forging at lower temperatures was successful eventually.

The outcomes of the austenitizing trials were similar for both materials, leading to a temperature of 950°C and a soaking time of 30 minutes which led to fully martensitic microstructure after quenching. Finally, 425°C was chosen as the intercritical annealing temperature – on the basis of the mechanical properties and microstructures obtained in the trials.

The findings from the manufacture of the stock and from the austenitizing and austempering trials will be important for future thermomechanical treatment of advanced multiphase high-strength aluminium-alloyed steels. The ultimate goal is to apply this knowledge to a general design of thermomechanical treatment of third-generation advanced high-strength steels in the form of an upgraded press-hardening process.

Acknowledgement

This paper presents results of Student Grant Competition at the University of West Bohemia, SGS 2019-019 – Materials and technological research into advanced high-strength steels, supported from specific-purpose funds of the state budget for research and development administered by the Ministry of Education of the Czech Republic.

References

[1] Karbasian H Tekkaya A E 2010 *A review on hot stamping*, Journal of Materials Processing Technology **210** 2103-18.
[2] Kučerová L, Jirková H and Mašek B 2014 *The Effect of Alloying on Mechanical Properties of Advanced High Strength Steels* Archives of Metallurgy and Materials **59** 1189-92.
[3] Tamarelli C M 2011 *AHSS 101: The Evolving Use of Advanced High-Strength Steels for Automotive Applications*, Southfield, Michigan.
[4] *Automotive Steels: Design, Metallurgy, Processing and Applications*, Woodhead Publishing Series in Metals and Surface Engineering, 2017 ISBN: 978-0-08-100638-2.
[5] Totten G E 2007 *Steel heat treatment: metallurgy and technologies*, 2nd edition, ISBN-13: 978-0-8493-8455-4.

[6] Bhadeshia H K D H 2001 Bainite in steels, 2nd edition, London: Institute of Materials ISBN 1-86125-112-2.

[7] SENTE SOFTWARE LTD 2017, JMatPro the Materials Property Simulation Package Version 9.0 United Kingdom.

[8] Kim H Inoue J Okada M Nagat K 2017 Prediction of Ac3 and Martensite start temperatures by a data-driven model selection approach, ISIJ International **57** No. 12 2229-36.

[9] Cooman B C 2004 Structure-properties relationship in TRIP steels containing carbide-free bainite, Current Opinion in Solid State and Materials Science 8 pp. 285-303.

[10] Jirková H Opatová K Jeníček, Š, Vrtáček J Kučerová, L Kurka P 2019. Use of multi-phase trip steel for press-hardening technology. *Acta Metallurgica Slovaca*, 25(2), 101-106. doi:10.12776/ams.v25i2.1267

[11] Kučerová L., Bystrianský M., Jeníček Š., Francisko, P 2017. Effect of deformation conditions on microstructure and mechanical properties of low alloyed steel. Manufacturing Technology, 17(5), 752-756.

[12] Jeníček Š, Bublíková D, Jirková H., Kána J. 2017. Stability of retained austenite in high-strength martensitic steels during cold deformation. Paper presented at the Annals of DAAAM and Proceedings of the International DAAAM Symposium, 289-294.

[13] Kotěšovec V, Vorel I., Jeníček Š., Káňa, J, brahim, K. 2017. Impact of quenching temperature and isothermal holding time during austempering on bainite content in high-silicon steel. Paper presented at the *IOP Conference Series: Materials Science and Engineering, , 179*(1) doi:10.1088/1757

[14] Kučerová L., Jeníček Š., Mach J. 2018. Effect of sample geometry on tensile properties of chromium alloyed middle carbon steel. Paper presented at the *METAL 2018 - 27th International Conference on Metallurgy and Materials, Conference Proceedings,* 761-765.

AUTHOR INDEX